T0332884

Continuous-Time Markov-Modulated Chains in Operations Research

Continuous-Time Markov-Modulated Chains in Operations Research

Alexander Andronov
Kristina Mahareva
Transport and Telecommunication Institute (TSI), Latvia

NEW JERSEY • LONDON • SINGAPORE • BEIJING • SHANGHAI • HONG KONG • TAIPEI • CHENNAI • TOKYO

Published by

World Scientific Publishing Co. Pte. Ltd.
5 Toh Tuck Link, Singapore 596224
USA office: 27 Warren Street, Suite 401-402, Hackensack, NJ 07601
UK office: 57 Shelton Street, Covent Garden, London WC2H 9HE

Library of Congress Cataloging-in-Publication Data
Names: Andronov, Alexander, 1937- author. | Mahareva, Kristina, author.
Title: Continuous-time Markov-modulated chains in operations research /
Alexander Andronov, Transport and Telecommunication Institute (TSI), Latvia, Kristina Mahareva, Transport and Telecommunication Institute (TSI), Latvia.
Description: New Jersey : World Scientific, [2024] | Includes bibliographical references.
Identifiers: LCCN 2024000390 | ISBN 9789811286155 (hardcover) |
ISBN 9789811286162 (ebook) | ISBN 9789811286179 (ebook other)
Subjects: LCSH: Markov processes. | Poisson processes. | Operations research.
Classification: LCC QA274.7 .A56 2024 | DDC 519.2/33--dc23/eng/20240310
LC record available at https://lccn.loc.gov/2024000390

British Library Cataloguing-in-Publication Data
A catalogue record for this book is available from the British Library.

For any available supplementary material, please visit
https://www.worldscientific.com/worldscibooks/10.1142/13675#t=suppl

Desk Editors: Nimal Koliyat/Tan Rok Ting

Typeset by Stallion Press
Email: enquiries@stallionpress.com

Printed in Singapore

To the memory of Professor
Khaim Borisovich Kordonsky,
Educator and Researcher

About the Authors

Alexander Andronov, born in 1937, is a distinguished Latvian scientist and Emeritus Professor at the Transport and Telecommunication Institute. He completed his education at the Aviation Technological Institute and dedicated his entire career to working in prestigious universities: Riga Aviation University as Head of Department of Computer Systems for Information Processing and Management, and Riga Technical University as the Head of Division of Mathematical Support of Management Computer Systems for Transport and Transport and Telecommunication Institute. In 2021, he was honored with the prestigious title of Emeritus Professor at the Transport and Telecommunication Institute in Riga, Latvia.

With a doctorate in Engineering Sciences, Professor Andronov's unwavering passion for research and scientific endeavors has earned him the recognition of the Republic of Latvia, which awarded him the title of Honorary Scientist in 1990.

Throughout his career, Professor Andronov has been actively involved in research and has authored more than 100 influential publications in the field of mathematics. His research papers have been widely recognized and have contributed to advancements in mathematical theory and applications. Professor Andronov's expertise and mentorship have also nurtured the talents of aspiring mathematicians, shaping the future of the discipline and paving the way for more than 30 students to achieve their doctoral degrees.

As a testament to his expertise, Professor Andronov serves on the editorial board of the journal *Automatic Control and Computer Sciences* published by Springer. He has also been a member of the American Statistical Association (ASA). Professor Andronov's contributions to the field of mathematics, research, and teaching have left an indelible mark on the scientific community.

Kristina Mahareva, FCCA, CIA, is a prominent Latvian economist and researcher in statistical finance. With a Ph.D. in Economics awarded in 2020 in Latvia, she is actively involved in various research programs that focus on mathematical and statistical research, as well as real business applications to solve complex problems. Her research includes the development of methodology for aggregator models, AML analytics (Anti-Money Laundering modeling), auditing compliance, and applications in industries such as finance, travel and tourism, and the sustainable power supply industry.

Dr. Mahareva's passion for her field shines through in her numerous publications and her dedicated supervision of Ph.D. students. Beyond her academic work, she has organized and facilitated programs and workshops in Latin America, aimed at inspiring and empowering women in Science. Dr. Mahareva continues to make a profound impact, inspiring colleagues and students alike as she advances the boundaries of economics and research.

Contents

Introduction

Probabilistic models are widely used to describe and analyze various processes in system reliability, risk, queuing, communication, logistics, storage systems, etc. A significant place here belongs to Markovian models, introduced by the Russian mathematician A.A. Markov [1,2]. It is a *Markov chain* if values of the process form a finite or denumerable set; otherwise, it is a Markov process. This distinguishes between *continuous-time* and *discrete-time* models. There are a number of prominent books on Markov models [3–9].

Recently, these models have been considered with the appreciation of an external random environment [10,11]. The latter is presented as a continuous-time Markov chain. This bivariate process is called a *Markov-modulated process* and is denoted by (J, X). Here, X is a *basic component,* and J is an external random environment called the *Markov component.* Generally, the parameters of the distribution of component X are dependent on the state of component J. If process J is non-observable, some authors have discussed *Hidden Markov models* [10]. The terms *Markov* and *regime switching* are often used.

Our book is devoted to various applications of the Markov modulated processes theory. This book includes 11 chapters. A brief summary and chapter descriptions are provided below.

The book begins with Chapter 1: *Continuous-Time Markov Chains.* This chapter introduces the basic concepts of continuous-time Markov chains. It also presents a formula for the non-stationary probabilities of states for continuous-time finite Markov chains. This is consistently used within the framework of the entire book.

Chapter 2 is devoted to defining the *Markov-Modulated Chains.* For example, the famous machine repair problem is considered as an example of a random external environment existence. Erlang models are examined for the same case, and results are presented.

Chapter 3 describes the *Shortest Paths in Markov-Modulated Networks.* The problem of the shortest path is a classical problem in graph theory. We consider one case in which the travel speeds on the arcs are random variables, depending on the state of the external environment.

Chapter 4 describes the *Network Flows with Markov-Modulated Costs of Arcs.* It is supposed that a flow is set. It is necessary to compute the probabilistic distribution of the flow cost in the time interval $(0, t)$ for the stationary and non-stationary cases.

Chapter 5, *Wear Process Modulated by Cycling Continuous-Time Markov Chains*, contains an example from reliability theory. It is considered a case of gradual failure. If the Markov chain is in the ith state during time t, then the wear value increases on $c_i t$. The process ends when the accumulated wear reaches a critical level. Our goal is to compute the time distribution until it reaches the critical level. The renewal process is also considered.

In Chapter 6, we investigate the Poisson process, which is considered to be a flow of arrivals. It is assumed that the process has varying intensities generated by a continuous-time alternating Markov chain. The following indices of the flow are considered: the distribution, expectation, variance of the number of arrivals within a given interval, and the correlation of the number of arrivals between two adjacent intervals. The final section of the chapter is devoted to statistical estimation of the process parameters.

In Chapter 7, the *Markov-Modulated Poisson Process* is considered when a random environment has an arbitrary finite number of states. The results obtained were used to solve an inventory problem.

In Chapter 8, we consider a situation in which one process $X(t)$ or $J(t)$ is fully observed, and we wish to determine the state of the other process.

Monitoring is the process of observing an object with the purpose of controlling its state. Often, indicators of interest cannot be observed directly, and it is necessary to use other indicators that are connected with indicators of interest. In this book, the main probabilistic processes $X(t)$ are immersed in an external random environment, represented as a continuous-time finite Markov chain $J(t)$.

In Chapter 9, the popular reliability model k-out-of-n is the focus. Numerous studies have previously been conducted on this problem. For the first time, we consider the existence of an external random environment.

In Chapter 10, statistical problems of parameter estimation for the continuous-time Markov chain are considered, which describes the external random environment. The distribution of estimates was obtained. This allowed us to make various statistical inferences regarding the parameters.

Chapter 11 reviews other applications of continuous-time Markov-modulated processes in operational research.

Numerous examples in the chapters are accompanied by formal descriptions. All the numerical computations were performed using Mathcad. This allowed us to successfully overcome all computation difficulties. First, it calculates the transient probabilities of the states for a continuous-time finite Markov chain. These probabilities are expressed using the eigenvalues and eigenvectors of the corresponding matrix (generator). This process is called *spectral decomposition.*

A simple explicit form of the solution is missing in a more complex case of differential or integral equations. The explicit form of the solution is presented using infinite sums of functions. For example, we must often deal with the *renewal equation* [6]. Its solution is presented as an infinite sum of *the renewal function.* In this case, an approximation of the functions of interest and iterative computation procedures are used.

To read the book, some knowledge of probability theory and matrix algebra is necessary. The book is intended for specialists in different areas interested in the mathematical modeling of various production processes. It can be recommended as a manual for master's and PhD students.

The book adds to a library of applications of Markov-modulated processes.

Some chapters of the book were written with the help of Nadezhda Spiridovska and Diana Santalova (Chapter 3) and Irina Yackiva and Diana Santalova (Chapter 10). We extend our heartfelt appreciation to Springer Nature for graciously allowing us to repurpose previously published materials within Chapter 6 and Chapter 10.

References

[1] Markov, A.A., Rasprostranenie zakona bol'shih chisel na velichiny, zavisyaschie drug ot druga. *Izvestiya Fiziko-matematicheskogo obschestva pri Kazanskom universitete*, 2-ya seriya, tom 15, 1906: 135–156.

[2] Markov, A.A., Extension of the limit theorems of probability theory to a sum of variables connected in a chain. [Reprinted in Appendix B of R. Howard.] *Dynamic Probabilistic Systems, volume 1: Markov Chains.* John Wiley and Sons, 1971.

[3] Anderson, W.J., *Continuous-Time Markov Chains: An Applications-Oriented Approach.* Springer-Verlag. New York, Berlin, Heidelberg, London, Paris, Tokyo, Hong Kong, Barcelona, 1991.

[4] Bharucha-Reid, A.T., *Elements of the Theory of Markov Processes and their Applications.* McGraw-Hill, New York, 1960.

[5] Çinlar, E., *Introduction to Stochastic Processes.* Prentice-Hall, New Jersey, 1975.

[6] Feller, W., *An Introduction to Probability Theory and its Applications,* Vol. II. John Wiley and Sons, Inc., New York, London, Sydney, Toronto, 1971.

[7] Kemeny, J.G., Snell, J.L., *Finite Markov Chains.* Van Nostrand Reinhold Company, New York, 1960.

[8] Kemeny, J.G., Snell, J.L., Knapp, A.W., *Denumerable Markov Chains.* Springer-Verlag, New York, Heidelberg, Berlin, 1976.

[9] Kijima, M., Markov Processes for Stochastic Modeling, Cambridge, UK: The University Press, 1997.

[10] Mamon, R.S., Elliot, R.J. (eds.), *Hidden Markov Models in Finance, International Series in Operations Research & Management Science,* Vol. 104. Springer, New York, 2007.

[11] Pacheco, A., Tang, L.C., Prabhu, N.U., *Markov-Modulated Processes & Semi regenerative Phenomena.* World Scientific, New Jersey, London, 2009.

Chapter 1

Continuous-Time Markov Chains

1.1 Solution of a system of first-order differential equations with constant coefficients

In this book, the stochastic process $J(t)$, $t \geq 0$, is considered constantly. Argument t is continuous time. The values of process $J(t)$ belong to a finite set of natural numbers $\{1, 2, \ldots, k\}$ where the elements are called *the states*. The sojourn time of process $J(t)$ in the ith state does not depend on the prehistory and has an exponential distribution with the parameter $\Lambda_i > 0$. Then, a transition to the other state occurs and so on.

The probability $P\{J(t) = j | J(0) = i\}$ of the transition from state i to state j at time t is of great interest. Furthermore, we will see that these transient probabilities are the solution of a system of differential equations with a constant coefficient. Below we present the necessary knowledge for the solution of such systems [1,12].

Let us consider a system that has k unknown functions $x_i(t)$ of time $t \geq 0$ and k differential equations:

$$\frac{d}{dt}x_i(t) = \sum_{j=1}^{k} A_{i,j}x_j(t), x_i(0) = c_i, \quad i = 1, \ldots, k,$$

where $\{A_{i,j}\}$ and $\{c_i\}$ are known constants.

Using vector–matrix denotation $c = (c_1, \ldots, c_k)^T$, $x(t) = (x_1(t), \ldots, x_k(t))^T$, $A = (A_{i,j})_{k \times k}$, we get an equation

$$\frac{d}{dt}x(t) = Ax(t), \quad x(0) = c. \tag{1.1}$$

A solution of the differential equations system (1.1) is given through eigenvalues $\chi_1, \ldots, \chi_k$ and eigenvectors $\beta_1, \ldots, \beta_k$ of a square matrix A (see [1,12]).

Note that the scalar χ_* is called *the eigenvalue* of matrix A, corresponding to an eigenvector β_*, if

$$A\beta_* = \chi_* \beta_*. \tag{1.2}$$

It turns out that the number of eigenvalues equals the matrix dimension. It follows from Definition (1.2) that the eigenvectors are determined up to a scalar multiplier. Usually, it is chosen such that the norm of the vectors equals the unit:

$$\beta_*^T \beta_* = 1. \tag{1.3}$$

Eigenvalues and eigenvectors have been calculated using many computer programs. We consider the most frequently occurring practical case when all eigenvalues are different. We denote $\chi_1, \ldots, \chi_k$, and a vector with these components $\chi = (\chi_1, \ldots, \chi_k)^T$. Furthermore, $\text{diag}(\chi)$ is a diagonal matrix with diagonal $\chi = (\chi_1, \ldots, \chi_k)$.

Let B be a matrix whose columns are eigenvectors: $B = (\beta_1, \ldots, \beta_k)$. Let $\tilde{\beta}_1, \ldots, \tilde{\beta}_k$ be rows of the inverse matrix, so $B^{-1} = (\tilde{\beta}_1^T, \ldots, \tilde{\beta}_k^T)^T$.

Formula (1.1) shows that the solution depends on the initial condition $x(0) = c$. This solution can be written in a simpler way if k initial conditions are considered simultaneously, and the ith initial condition is as follows: the ith component equals 1, and the other components are equal to zero. Now, instead of one vector of the initial conditions, we have a matrix, which is a unit matrix I.

Let $X(t)$ be a matrix whose vector-columns are the solutions for different initial conditions. Now the system (1.1) for the considered initial conditions is represented by a matrix form:

$$\frac{d}{dt} X(t) = \mathbf{A} X(t), \quad t \geq 0. \tag{1.4}$$

The matrix form of Eq. (1.2) is the following:

$$AB = B \, \text{diag}(\chi). \tag{1.5}$$

Theorem 1.1. *A solution for the considered system of differential equations for different conditions is given as*

$$X(t) = \sum_{i=1}^{k} \exp(\chi_i t)\beta_i\tilde{\beta}_i = B\,\mathrm{diag}(\exp(\chi t))B^{-1}, \quad t \geq 0, \qquad (1.6)$$

where $\exp(\chi t) = (\exp(\chi_i t) \quad \cdots \quad \exp(\chi_k t))^{\mathrm{T}}$.

Proof. First, we show that the last expression suffices (1.4). Using formula (1.2), we obtain

$$\begin{aligned}\frac{d}{dt}X(t) &= \frac{d}{dt}\sum_{i=1}^{k} \exp(\chi_i t)\beta_i\tilde{\beta}_i = \sum_{i=1}^{k} \exp(\chi_i t)\chi_i\beta_i\tilde{\beta}_i \\ &= \sum_{i=1}^{k} \exp(\chi_i t)A\beta_i\tilde{\beta}_i = A\sum_{i=1}^{k} \exp(\chi_i t)\beta_i\tilde{\beta}_i = AX(t).\end{aligned}$$

Secondly, we must demonstrate that the initial condition $X(0) = I$, where I is a unit matrix, is satisfied. For this, we use the following alternative of a product calculation for two square matrices B and C of the same dimension k. If β_i is a column of matrix B and $\tilde{c}_i$ is a row of matrix C, then

$$BC = \sum_{i=1}^{k} \beta_i\tilde{c}_i.$$

Now the requisite follows from the equations:

$$X(0) = \sum_{i=1}^{k} \exp(\chi_i 0)\beta_i\tilde{\beta}_i = \sum_{i=1}^{k} \beta_i\tilde{\beta}_i = BB^{-1} = I. \qquad \square$$

Example 1.1. Consider a system of three components with the following matrix of constant coefficients:

$$\mathrm{A} = \begin{pmatrix} 0 & 1 & 0 \\ 0.06 & 0 & 0.01 \\ -0.005 & -0.005 & 0 \end{pmatrix}.$$

A computer calculation yields the following values for a vector of eigenvalues $\chi = (\chi_1, \ldots, \chi_k)^T$ and a matrix of eigenvectors

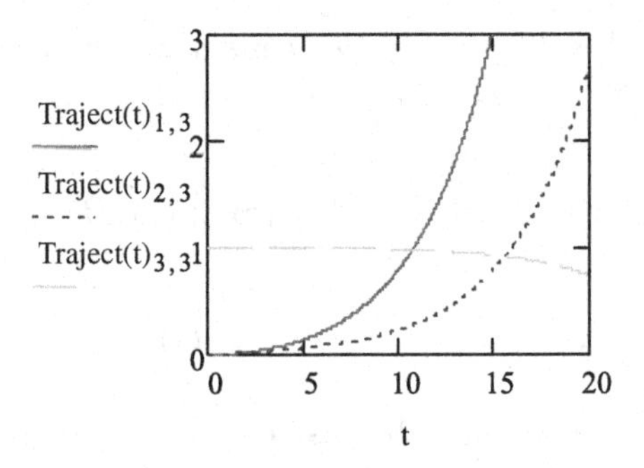

Fig. 1.1. Graphic of component state changes.

$B = (\beta_1, \ldots, \beta_k)$:

$$\chi = \begin{pmatrix} -0.245 \\ 0.244 \\ 8.34 \times 10^{-4} \end{pmatrix}, \quad B = \begin{pmatrix} 0.971 & -0.971 & -0.164 \\ -0.238 & -0.237 & -1.371 \times 10^{-4} \\ 0.015 & 0.025 & 0.986 \end{pmatrix}.$$

The initial state of the system components is the following: $c = (0, 0, 1)^T$.

Figure 1.1 shows how the states of the component change at time t, where $\text{Traject}(t)_{i,j} = X(t)_{i,j}$.

1.2 Continuous-time finite Markov chain

This is a stochastic process $J(t)$, $t \geq 0$, with k state, $k < \infty$ (see, for example, [2–10]). The continuous-time Markov chain is determined by the non-negative parameters $\lambda_{i,j} \geq 0$, $i, j = 1, \ldots, k$. It is assumed that the sojourn time in state i has an exponential distribution with parameter $\Lambda_i = \sum_{j=1}^{k} \lambda_{i,j}$. When the sojourn time in state i ends, the chain transfers in state j with probability $\lambda_{i,j}/\Lambda_i$.

Often, the *irreducible* Markov chain is considered. This means that "its only closed set of states is the set of all states" (see [6], p. 32).

Now, value $\lambda_{i,j}\Delta t$ can be interpreted as the probability that the chain transitions from state i to state j within small time interval Δt.

We have

$$P_{i,j}(\Delta t) = P\{J(\Delta t) = j|J(0) = i\}$$
$$= \Lambda_i \Delta t \frac{\lambda_{i,j}}{\Lambda_i} + o(\Delta t) = \lambda_{i,j}\Delta t + o(\Delta t).$$

We wish to determine the probability that the chain will be in state j at time t if the initial state is i:

$$P_{i,j}(t) = P\{J(t) = j|J(0) = i\}, \quad t \geq 0.$$

The usual reasoning for a short time interval Δt leads to the following equation if the first jump after the initial time moment 0 is considered:

$$P_{i,j}(t + \Delta t) = (1 - \Delta t \Lambda_i) P_{i,j}(t)$$
$$+ \sum_{\nu=1}^{k} \lambda_{i,v} \Delta t P_{v,j}(t) + o(\Delta t), \quad t \geq 0, \ \forall i, j.$$

Further, the equation

$$\frac{1}{\Delta}(P_{i,j}(t + \Delta) - P_{i,j}(t)) = -\Lambda_i P_{i,j}(t) + \sum_{\nu=1}^{k} \lambda_{i,v} P_{v,j}(t)$$
$$+ \frac{o(\Delta t)}{\Delta t}, \quad t \geq 0, \ \forall i, j,$$

leads us to a system of differential equations with constant coefficients:

$$\dot{P}_{i,j}(t) = -\Lambda_i P_{i,j}(t) + \sum_{\nu=1}^{k} \lambda_{i,v} P_{v,j}(t), \quad t \geq 0, \ \forall i, j. \tag{1.7}$$

Next, we introduce the following notations: $P(t) = (P_{i,j}(t))_{k \times k}$ and $\lambda = (\lambda_{i,j})_{k \times k}$ are matrices, $\Lambda = (\Lambda_1, \Lambda_2, \ldots, \Lambda_k)^T$ is a vector, and $\text{diag}(\Lambda)$ is a diagonal matrix with vector Λ in the main diagonal. Now, we rewrite system (1.7) in matrix form as

$$\dot{P}(t) = -\text{diag}(\Lambda) P(t) + \lambda P(t),$$
$$\dot{P}(t) = (\lambda - \text{diag}(\Lambda)) P(t), \quad t \geq 0. \tag{1.8}$$

This system is known as the *backward Kolmogorov equation*.

We previously considered system (1.4), where $X(t) = P(t)$, $A = \lambda - \text{diag}(\Lambda)$. The $(k \times k)$ matrix $A = \lambda - \text{diag}(\Lambda)$ is called the *generator* of the Markov chain. It should be recalled that eigenvalues and eigenvectors of this matrix are denoted by $\chi_1, \chi_2, \ldots, \chi_k$ and $\beta_1, \beta_2, \ldots, \beta_k$, respectively. We consider a case in which all values $\chi_1, \chi_2, \ldots, \chi_k$ are different, one eigenvalue equals zero, and the other eigenvalues are negative. This zero eigenvalue will be denoted as χ_1.

The matrix, whose columns are eigenvalues, is denoted $B = (\beta_1, \ldots, \beta_k)$, and its inverse matrix with rows $\tilde{\beta}_1, \ldots, \tilde{\beta}_k$ is denoted $B^{-1} = \tilde{B} = (\tilde{\beta}_1^T, \ldots, \tilde{\beta}_k^T)^T$.

Now it follows from Theorem 1.1 that

$$P(t) = \sum_{i=1}^{k} \exp(\chi_i t)\beta_i\tilde{\beta}_i = B\,\text{diag}(\exp(\chi t))B^{-1}, \quad t \geq 0, \tag{1.9}$$

where $\exp(\chi t) = (\exp(\chi_i t) \quad \cdots \quad \exp(\chi_k t))^T$.

This formula allows the calculation of the non-stationary probabilities of states $P_{i,j}(t) = P\{J(t) = j|J(0) = i\}$. Furthermore, it is possible to calculate $ET_{i,\nu}(t)$ — the average value of the sojourn time in state v on the interval $(0, t)$, if the initial state of the chain is i. It is possible to use a numerical integration or an analytical formula

$$\begin{aligned} ET(t)_v &= \int_0^t P_{i,v}(u)du = \int_0^t \sum_{\eta=1}^{k} B_{i,\eta} \exp(\chi_\eta u)\bar{B}_{\eta,v}du \\ &= tB_{i,1}\bar{B}_{1,v} + \sum_{\eta=2}^{k} B_{i,\eta} \int_0^t \exp(\chi_\eta u)du\bar{B}_{\eta,v} \\ &= tB_{i,1}\bar{B}_{1,v} + \sum_{\eta=2}^{k} B_{i,\eta}\frac{1}{\chi_\eta}(\exp(\chi_\eta t) - 1)\bar{B}_{\eta,v}, \quad v = 1, \ldots, k. \end{aligned} \tag{1.10}$$

It is said that a Markov chain has a *stationary distribution of a state's probabilities* if the following limits do not depend on the initial states of the chain:

$$\begin{gathered} \lim_{t\to\infty} P_{i,j}(t) = p_j, \quad \forall j\, p_j \geq 0, \\ \sum_j p_j = 1. \end{gathered} \tag{1.11}$$

It turns out that for the irreducible chain, only one eigenvalue of matrix $A = \lambda - \mathrm{diag}(\Lambda)$ equals zero, and the other values are negative. Furthermore, all components of the eigenvector corresponding to the zero eigenvalue are identical. For simplicity, let us assume that the first eigenvalue is zero. Subsequently, with respect to formula (1.9),

$$\lim_{t\to\infty} P(t) = \lim_{t\to\infty} \sum_{i=1}^{k} \exp(\chi_i t)\beta_i\tilde{\beta}_i$$

$$= \exp(0)\beta_1\tilde{\beta}_1 + \lim_{t\to\infty} \sum_{i=2}^{k} \exp(\chi_i t)\beta_i\tilde{\beta}_i = \beta_1\tilde{\beta}_1. \qquad (1.12)$$

Because all components of vector β_1 are the same, all rows of the matrix $\beta_1\tilde{\beta}_1$ are the same. Each row provides a vector of stationary probabilities of the states $p = (p_1, \ldots, p_k)$.

The obtained results are presented in the following theorem.

Theorem 1.2. *Non-stationary probabilities of the states are given in formula* (1.9). *The average value of the sojourn time in the states on the interval* $(0, t)$ *is calculated using formula* (1.10). *The vector of stationary probabilities of the states is given in formula* (1.12).

Stationary probabilities $p = (p_1 \cdots p_k)$ exhibit the known property:

$$pP(t) = p, \quad t \geq 0.$$

Let us prove this statement. We know that

$$\begin{pmatrix} p \\ \cdots \\ p \end{pmatrix} = \beta_1\tilde{\beta}_1.$$

Therefore,

$$\begin{pmatrix} p \\ \cdots \\ p \end{pmatrix} P(t) = \beta_1\tilde{\beta}_1 P(t) = \sum_{i=1}^{k} \exp(\chi_i t)\beta_1\tilde{\beta}_1\beta_i\tilde{\beta}_i$$

$$= \exp(\chi_1 t)\beta_1\tilde{\beta}_1 = \beta_1\tilde{\beta}_1 = \begin{pmatrix} p \\ \cdots \\ p \end{pmatrix},$$

because $\chi_1 = 0$, $\tilde{\beta}_1\beta_1 = 1$, $\tilde{\beta}_1\beta_i = 0$, $i \neq 1$.

This ends the proof.

Mitrani first draws attention to the presented results in the paper "The spectral extension solution method for Markov processes on lattice strips" [11].

Example 1.2 (Birth-death process). This process has been repeatedly considered in the literature [1,2,5,10–12]. It is usually described as follows. The states of the process $J(t)$ are integer non-negative variables. We confine ourselves to the case in which the number of states is finite, so $J(t) \in \{0, \dots, k-1\}$, $k < \infty$. A transition from each state to a neighboring state is possible. Let α_i be the transition intensity from state i to state $i+1$ (birth intensity) and β_i – to state $i-1$ (death intensity), $\alpha_{k-1} = \beta_0 = 0$.

Here, the above Markov chain takes place with the following non-zero transient intensities:

$$\lambda_{i,i+1} = \alpha_i, \quad i = 0, \dots, k-2,$$

$$\lambda_{i,i-1} = \beta_i, \quad i = 1, \dots, k-1.$$

The theory presented above allows the calculation of the non-stationary distribution of state probabilities.

As an example, let us consider a well-known problem, the *machine repair problem*, for the following input data: the number of machines is equal to 4. The machines fail independently, each with a constant intensity of 1. The repair of the machine requires one worker. The repair intensity for one worker equals 2. There are two workers. If the number of failed machines exceeds two, then the unserviced machines are idle and wait for repair. We wish to calculate the probability distribution for a number of failed machines $X(t)$ as a function of time t. Let us assume that all the machines are in good working order at the initial time $t = 0$.

We have five states. Furthermore, $J(t) = i$ means that i machines are in good working order at t. Here, α_i is the total intensity of machine failure, β_i is the total intensity of the repair.

An intensity matrix $\lambda = (\lambda_{i,j})$ and generator A are the following:

$$\lambda = \begin{pmatrix} 0 & 4 & 0 & 0 & 0 \\ 2 & 0 & 3 & 0 & 0 \\ 0 & 4 & 0 & 2 & 0 \\ 0 & 0 & 4 & 0 & 1 \\ 0 & 0 & 0 & 4 & 0 \end{pmatrix}, \quad A = \begin{pmatrix} -4 & 4 & 0 & 0 & 0 \\ 2 & -5 & 3 & 0 & 0 \\ 0 & 4 & -6 & 2 & 0 \\ 0 & 0 & 4 & -5 & 1 \\ 0 & 0 & 0 & 4 & -4 \end{pmatrix}.$$

The eigenvalues vector χ, the matrix of the eigenvectors B and its inverse matrix B^{-1} are presented as follows:

$$\chi = \begin{pmatrix} -10.476 \\ -6.906 \\ 0 \\ -4.461 \\ -2.156 \end{pmatrix}, \quad B = \begin{pmatrix} 0.27 & -0.343 & 0.447 & -0.348 & -0.290 \\ -0.436 & 0.249 & 0.447 & 0.040 & -0.134 \\ 0.617 & 0.070 & 0.447 & 0.239 & 0.067 \\ -0.508 & -0.531 & 0.447 & 0.104 & 0.396 \\ 0.314 & 0.730 & 0.447 & -0.900 & 0.858 \end{pmatrix},$$

$$B^{-1} = \begin{pmatrix} 0.218 & -0.706 & 0.748 & -0.308 & 0.048 \\ -0.612 & 0.889 & 0.188 & -0.710 & 0.244 \\ 0.411 & 0.822 & 0.617 & 0.308 & 0.077 \\ -0.941 & 0.217 & 0.970 & 0.210 & -0.457 \\ -0.759 & -0.70 & 0.262 & 0.776 & 0.421 \end{pmatrix}.$$

This allows us to calculate the transition probabilities $P_{0,j}(\tau)$ for different times τ using formula (1.9). The results are shown in Fig. 1.2.

Let us calculate the stationary probabilities of the states. We see that the zero eigenvalue has the number 3 in the vector χ. By multiplying the third column of matrix B with the third row of matrix B^{-1}, we obtain the matrix

$$P(\infty) = \begin{pmatrix} 0.184 & 0.368 & 0.276 & 0.138 & 0.034 \\ \dots & \dots & \dots & \dots & \dots \\ 0.184 & 0.368 & 0.276 & 0.138 & 0.034 \end{pmatrix}.$$

The rows of this matrix give a stationary distribution of the state probabilities.

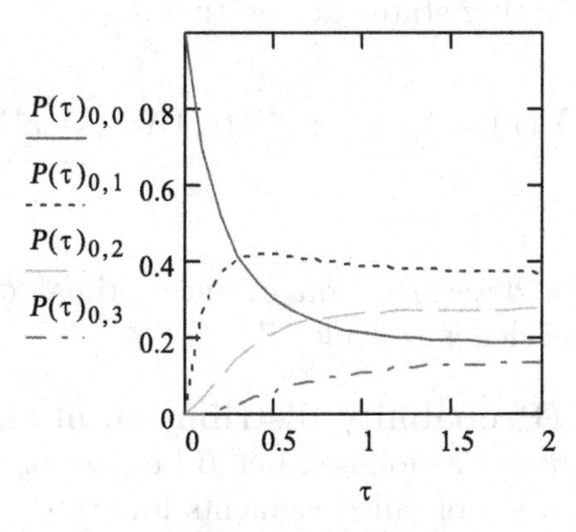

Fig. 1.2. Non-stationary probabilities of the states.

Example 1.3 (Reliability of renewable elements' system). The considered system contains n elements. These elements fail independently. The working time of the ith element has an exponential distribution with intensity λ_i, $i = 1, \ldots, n$. There exist s servers, whose repair of failed elements is $s < n$. The absolute priorities of the renewal are considered. They are determined by a permutation ord of element numbers $1, 2, \ldots, n$. The value ord_i gives a priority number to the ith element. The highest priority is element number 1.

A failed element by number i is served if there exists a free server or if some element with a bigger priority number is served. In the latter case, the new failed element takes the place of the served one with the biggest priority number. An element with interrupted service waits for the continuation of the service. The renewal time of the ith element has an exponential distribution with intensity μ_i, $i = 1, \ldots, n$.

The state of the entire system is described by the Boolean n-vector $x = (x_1, \ldots, x_n)$. The value $x_i = 0$ indicates that the ith element fails, and the value $x_i = 1$ indicates that the ith element works. The Boolean function for n arguments $y(x)$ is known. A value of 0 indicates that the system fails; otherwise, it works.

The state of the system changes randomly, so we use the notation $X(t)$ for time t. Analogously, $Y(t) = y(X(t))$ is a system state at time t.

We consider the following tasks: first, the transition probabilities of the states $Pr_{x,x^*}(t) = P\{X(t) = x^*|X(0) = x\}$ are deduced. Second, we demonstrate the calculation of the reliability function $R_{x,x^*}(t)$. This is the probability that the considered process does not leave the set of working states during time t:

$$R_{x,x^*}(t) = P\{(Y(\tau) = 1 : 0 \leq \tau \leq t) \wedge (X(t) = x^*)|X(0) = x\}. \tag{1.13}$$

Furthermore, a determination of the optimal discipline of the failed elements' service is considered.

Example 1.3.1 (Probability distribution of the states). The required definitions are as follows. Let B be a set of all 2^n-dimension Boolean vectors. A set of failed elements for state $x \in B$ is denoted as $F(x)$:

$$F(x) = \{i : x_i = 0\}.$$

Let inord(j) be the number of elements that have jth priority. Note that ord_i and inord(j) are mutually inverse functions: $\text{inord}(\text{ord}_i) = i$. The set of serviced elements for state x is as follows:

$$S(x) = \left\{ i \in F(x) : \sum_{j=1}^{\text{ord}_i} (1 - x_{\text{inord}(j)}) \leq s \right\}, \quad x \in B.$$

Now, we express the transition intensities for the stochastic process $X(t)$. Let $N(x)$ be a set of states that differ in state x by only one component, and $i(x, x^*)$ be a serial number of different components for states x and $x^* \in N(x)$. Then, the transition intensity from state x to state $x^* \in N(x)$ is as follows [2,11]:

$$v(x, x^*) = \begin{bmatrix} \lambda_{i(x,x^*)} & \text{if } x_{i(x,x^*)} = 1, \\ \mu_{i(x,x^*)} & \text{if } i(x, x^*) \in S(x). \end{bmatrix}$$

The corresponding $2^n \times 2^n$-matrix of the transition intensities is denoted by V.

Let $\Delta(2^n)$ be the 2^n-column vector of the units and $d = V\Delta(2^n)$ be a column vector that contains sums of the elements of matrix V rows. A *generator* is a $2^n \times 2^n$-matrix

$$G = V - \text{diag}(d), \tag{1.14}$$

where diag(d) is a diagonal matrix with vector d on the main diagonal.

The non-stationary probabilities of the states are expressed using the generator G. To simplify the formulas, we use *decimal-coded binary* notations. This means that, instead of the binary vector x, its decimal value is used. Now, the notation $G_{i,j}$ is used instead of G_{x,x^*}.

Let χ be a vector of eigenvalues of generator G, and $\tilde{B}$ be a matrix of the corresponding normalized column-eigenvectors. As before, we suppose that all eigenvalues are different. The matrix of transition probabilities between the states is calculated by formula (1.9), and the stationary distribution of the state probabilities by formula (1.12).

Example 1.3.2 (Reliability function). Let $\Omega = \{x \in B : y(x) = 1\}$ be a set of working states, and m be the number of its elements. All the other states $B - \Omega$ form one absorbing state. We must calculate the $m \times m$-matrix $R(t) = (R_{x,x^*}(t))$ of the probabilities that our process $X(t)$ does not input in the absorbing state during time t.

Let $\tilde{V}$ be an $m \times m$-submatrix of matrix V, which has columns and rows from set Ω and $\tilde{d}$ be an analogous m-sub-vector of vector d. Now, the generator (1.14) is of the following form:

$$\tilde{G} = \tilde{V} - \text{diag}(\tilde{d}).$$

Analogous to the previous case, let $\tilde{\chi}$ be a vector of eigenvalues of the generator $\tilde{G}$, and $\tilde{Z}$ be a matrix of the corresponding normalized column-eigenvectors. Then

$$R(t) = (R_{i,j}(t)) = \sum_{k=1}^{m} \exp(t\tilde{\chi}_k)\tilde{Z}^{(k)}\tilde{\tilde{Z}}_{(k)}, \quad t \geq 0. \tag{1.15}$$

Further, the vector

$$R^*(t) = R(t)\Delta(2^m), \tag{1.16}$$

provides the reliability at time t for different initial states, irrespective of the final state $X(t)$.

Example 1.3.3 (Numerical results). A system of four elements is considered. The first and second elements are connected sequentially, and the third, fourth, and pair 1–2 are connected in parallel:

$$Y(t) = x(t)_1 \wedge x(t)_2 \vee x(t)_3 \vee x(t)_4.$$

This determines the following set of the working states of the system:

$$\Omega = \begin{pmatrix} 1 & 0 & 1 & 0 & 1 & 0 & 1 & 0 & 1 & 0 & 1 & 0 & 1 \\ 1 & 0 & 0 & 1 & 1 & 0 & 0 & 1 & 1 & 0 & 0 & 1 & 1 \\ 0 & 1 & 1 & 1 & 0 & 0 & 0 & 0 & 0 & 1 & 1 & 1 & 1 \\ 0 & 0 & 0 & 0 & 1 & 1 & 1 & 1 & 1 & 1 & 1 & 1 & 1 \end{pmatrix}$$

Table 1.1. The generator (1.14).

$$
G := \begin{pmatrix}
-10 & 1 & 2 & 0 & 3 & 0 & 0 & 0 & 4 & 0 & 0 & 0 & 0 & 0 & 0 & 0 \\
2 & -11 & 0 & 2 & 0 & 3 & 0 & 0 & 0 & 4 & 0 & 0 & 0 & 0 & 0 & 0 \\
3 & 0 & -11 & 1 & 0 & 0 & 3 & 0 & 0 & 0 & 4 & 0 & 0 & 0 & 0 & 0 \\
0 & 3 & 2 & -12 & 0 & 0 & 0 & 3 & 0 & 0 & 0 & 4 & 0 & 0 & 0 & 0 \\
0 & 0 & 0 & 0 & -7 & 1 & 2 & 0 & 0 & 0 & 0 & 0 & 4 & 0 & 0 & 0 \\
0 & 1 & 0 & 0 & 2 & -9 & 0 & 2 & 0 & 0 & 0 & 0 & 0 & 4 & 0 & 0 \\
0 & 0 & 1 & 0 & 3 & 0 & -9 & 1 & 0 & 0 & 0 & 0 & 0 & 0 & 4 & 0 \\
0 & 0 & 0 & 1 & 0 & 3 & 2 & -10 & 0 & 0 & 0 & 0 & 0 & 0 & 0 & 4 \\
0 & 0 & 0 & 0 & 0 & 0 & 0 & 0 & -6 & 1 & 2 & 0 & 3 & 0 & 0 & 0 \\
0 & 0 & 0 & 0 & 0 & 0 & 0 & 0 & 2 & -7 & 0 & 2 & 0 & 3 & 0 & 0 \\
0 & 0 & 0 & 0 & 0 & 0 & 0 & 0 & 3 & 0 & -7 & 1 & 0 & 0 & 3 & 0 \\
0 & 0 & 0 & 2 & 0 & 0 & 0 & 0 & 0 & 3 & 2 & -10 & 0 & 0 & 0 & 3 \\
0 & 0 & 0 & 0 & 0 & 0 & 0 & 0 & 0 & 0 & 0 & 0 & -3 & 1 & 2 & 0 \\
0 & 0 & 0 & 0 & 0 & 2 & 0 & 0 & 0 & 1 & 0 & 0 & 2 & -7 & 0 & 2 \\
0 & 0 & 0 & 0 & 0 & 0 & 2 & 0 & 0 & 0 & 1 & 0 & 3 & 0 & -7 & 1 \\
0 & 0 & 0 & 0 & 0 & 0 & 0 & 2 & 0 & 0 & 0 & 1 & 0 & 3 & 2 & -8
\end{pmatrix}
$$

The failure and repair intensities are $\lambda = (1 \quad 2 \quad 3 \quad 4)$, $\mu = (2 \quad 3 \quad 1 \quad 2)$. The number of s servers equals two. Initially, we consider absolute priorities that coincide with the element numbers: $\text{ord} = (1 \quad 2 \quad 3 \quad 4)$.

Generator G, calculated using formula (1.14), is presented in Table 1.1.

The vector of the eigenvalues χ is the following:

$$
\begin{aligned}
\chi := (&-17.175 - 1.275 \times 10^{-15} (-13.537 + 0.024\text{i}) \\
&\times (-13.537 - 0.024i) - 11.887 - 11.806 - 3 - 3.544 - 4.828 \\
&- 5 - 10.147 - 6.799 - 7.572 - 8.769 - 8 - 8.398).
\end{aligned}
$$

The matrix of eigenvectors Z is presented in Table 1.2. We can see that the zero value of the eigenvalue is -1.275×10^{-15}. The corresponding eigenvector has a constant value of 0.25 for its elements.

The matrix $\Pr(t)$ of the state probabilities for $t = 0.3$ is presented in Table 1.3.

We now consider the reliability of the system. The 13 working states are represented by matrix Ω. Further, the states $\{x_i\}$ are enumerated in accordance with matrix Ω. The generator used in this case is presented in Table 1.4. The corresponding eigenvalues are as

Table 1.2. Matrix of the eigenvectors of the generator (1.14).

$$
Z := \left(\begin{array}{cccccccc}
0.158 & 0.25 & 0.346 - 2.502i \times 10^{-3} & 0.346 + 2.502i \times 10^{-3} & -0.154 & -0.12 & -0.158 & 0.362 \\
-0.411 & 0.25 & 0.218 - 0.057i & 0.218 + 0.057i & 0.336 & 0.283 & 0.316 & 9.573 \times 10^{-3} \\
-0.27 & 0.25 & -0.574 & -0.574 & 0.242 & 0.239 & -0.158 & 0.468 \\
0.72 & 0.25 & -0.454 + 0.128i & -0.454 - 0.128i & -0.511 & -0.557 & 0.316 & 0.085 \\
-0.038 & 0.25 & -0.112 + 0.021i & -0.112 - 0.021i & 0.028 & 0.028 & -0.158 & -0.019 \\
0.153 & 0.25 & 0.038 - 0.038i & 0.038 + 0.038i & -0.105 & -0.083 & 0.316 & -0.287 \\
0.097 & 0.25 & 0.308 - 0.041i & 0.308 + 0.041i & -0.071 & -0.084 & -0.158 & 0.149 \\
-0.256 & 0.25 & -0.081 + 0.05i & -0.081 - 0.05i & 0.164 & 0.18 & 0.316 & -0.195 \\
-0.018 & 0.25 & 0.011 + 2.692i \times 10^{-3} & 0.011 - 2.692i \times 10^{-3} & -0.154 & -0.158 & -0.158 & 0.362 \\
0.08 & 0.25 & -0.112 + 3.204i \times 10^{-3} & -0.112 - 3.204i \times 10^{-3} & 0.336 & 0.344 & 0.316 & 9.573 \times 10^{-3} \\
0.046 & 0.25 & -0.013 - 3.083i \times 10^{-3} & -0.013 + 3.083i \times 10^{-3} & 0.242 & 0.243 & -0.158 & 0.468 \\
-0.296 & 0.25 & 0.358 - 0.047i & 0.358 + 0.047i & -0.511 & -0.494 & 0.316 & 0.085 \\
9.781 \times 10^{-3} & 0.25 & 0.019 - 5.689i \times 10^{-3} & 0.019 + 5.689i \times 10^{-3} & 0.028 & 0.029 & -0.158 & -0.019 \\
-0.063 & 0.25 & -1.159 \times 10^{-3} + 0.021i & -1.159 \times 10^{-3} - 0.021i & -0.105 & -0.116 & 0.316 & -0.287 \\
-0.038 & 0.25 & -0.102 + 0.019i & -0.102 - 0.019i & -0.071 & -0.067 & -0.158 & 0.149 \\
0.117 & 0.25 & 1.969 \times 10^{-3} - 0.028i & 1.969 \times 10^{-3} + 0.028i & 0.164 & 0.162 & 0.316 & -0.195
\end{array}\right.
$$

$$
\left.\begin{array}{cccccccc}
-0.073 & -0.196 & -0.323 & -0.217 & -0.261 & 0.291 & -0.124 & -0.397 \\
-0.137 & -0.196 & -0.471 & 0.329 & -0.106 & 0.201 & 0.248 & 0.271 \\
0.391 & 0.294 & 0.638 & -0.13 & -0.198 & -0.501 & 0.186 & 0.277 \\
0.317 & 0.294 & 0.267 & 0.537 & -0.659 & -0.237 & -0.372 & 0.151 \\
-0.133 & -0.196 & -0.155 & 0.035 & 0.046 & -1.972 \times 10^{-3} & -0.124 & -0.118 \\
-0.173 & -0.196 & 0.041 & -0.191 & 0.065 & -0.155 & 0.248 & 0.345 \\
0.35 & 0.294 & 0.184 & 0.029 & 0.175 & 0.083 & 0.186 & 0.041 \\
0.282 & 0.294 & 0.132 & -0.027 & -0.324 & 0.193 & -0.372 & -0.261 \\
-0.156 & -0.196 & -0.073 & -0.217 & -0.067 & 0.291 & -0.124 & -0.277 \\
-0.204 & -0.196 & -0.103 & 0.329 & 0.32 & 0.201 & 0.248 & 0.041 \\
0.316 & 0.294 & 0.174 & -0.13 & 0.059 & -0.501 & 0.186 & 0.41 \\
0.266 & 0.294 & 0.059 & 0.537 & -0.307 & -0.237 & -0.372 & -0.011 \\
-0.204 & -0.196 & 0.02 & 0.035 & -0.111 & -1.972 \times 10^{-3} & -0.124 & -0.065 \\
-0.221 & -0.196 & 0.118 & -0.191 & 0.189 & -0.155 & 0.248 & 0.173 \\
0.297 & 0.294 & -0.129 & 0.029 & 0.158 & 0.083 & 0.186 & 0.09 \\
0.24 & 0.294 & -0.194 & -0.027 & -0.168 & 0.193 & -0.372 & -0.421
\end{array}\right)
$$

Table 1.3. Matrix of state probabilities for $t = 0.3$.

Pr(0.3) :=

0.069	0.019	0.035	0.011	0.101	0.029	0.052	0.014
0.036	0.063	0.019	0.036	0.051	0.09	0.026	0.042
0.049	0.014	0.063	0.02	0.07	0.02	0.092	0.024
0.026	0.046	0.034	0.066	0.036	0.063	0.047	0.074
2.946×10^{-3}	5.125×10^{-3}	8.979×10^{-3}	3.273×10^{-3}	0.17	0.049	0.089	0.023
7.158×10^{-3}	0.025	6.348×10^{-3}	0.014	0.086	0.151	0.044	0.068
9.283×10^{-3}	4.781×10^{-3}	0.024	7.448×10^{-3}	0.118	0.033	0.154	0.038
6.436×10^{-3}	0.018	0.013	0.024	0.06	0.102	0.076	0.118
3.612×10^{-4}	1.254×10^{-3}	1.51×10^{-3}	3.796×10^{-3}	5.263×10^{-3}	0.011	0.019	6.102×10^{-3}
1.442×10^{-3}	6.553×10^{-3}	3.093×10^{-3}	0.016	9.78×10^{-3}	0.042	0.011	0.022
1.092×10^{-3}	2.352×10^{-3}	4.175×10^{-3}	8.355×10^{-3}	0.012	8.553×10^{-3}	0.041	0.012
5.753×10^{-3}	0.016	0.011	0.039	0.013	0.035	0.026	0.045
6.564×10^{-4}	1.609×10^{-3}	2.688×10^{-3}	1.58×10^{-3}	0.013	0.021	0.038	0.011
1.944×10^{-3}	9.891×10^{-3}	2.514×10^{-3}	7.642×10^{-3}	0.029	0.091	0.024	0.042
2.372×10^{-3}	1.901×10^{-3}	9.051×10^{-3}	4.12×10^{-3}	0.038	0.018	0.09	0.023
2.21×10^{-3}	8.14×10^{-3}	5.78×10^{-3}	0.015	0.025	0.063	0.046	0.074

0.159	0.043	0.078	0.018	0.224	0.045	0.084	0.019
0.081	0.135	0.038	0.053	0.104	0.129	0.04	0.057
0.111	0.029	0.139	0.03	0.144	0.03	0.135	0.032
0.055	0.087	0.066	0.087	0.067	0.085	0.065	0.095
5.91×10^{-3}	9.251×10^{-3}	0.016	4.683×10^{-3}	0.374	0.073	0.135	0.031
0.015	0.046	0.011	0.019	0.165	0.195	0.062	0.087
0.019	8.425×10^{-3}	0.045	0.01	0.227	0.046	0.205	0.049
0.012	0.03	0.022	0.032	0.106	0.131	0.1	0.149
0.228	0.061	0.112	0.025	0.319	0.064	0.117	0.027
0.116	0.191	0.054	0.074	0.146	0.177	0.055	0.077
0.159	0.04	0.197	0.042	0.202	0.041	0.186	0.043
0.075	0.118	0.089	0.115	0.091	0.113	0.086	0.124
8.199×10^{-3}	0.013	0.023	6.377×10^{-3}	0.531	0.101	0.186	0.042
0.02	0.061	0.015	0.025	0.222	0.255	0.082	0.113
0.026	0.011	0.06	0.014	0.308	0.061	0.269	0.065
0.017	0.04	0.029	0.041	0.14	0.171	0.13	0.193

Table 1.4. The generator $\tilde{G}$.

$$\begin{pmatrix}
-6 & 1 & 2 & 0 & 3 & 0 & 0 & 0 & 0 & 0 & 0 & 0 & 0 \\
2 & -7 & 0 & 2 & 0 & 3 & 0 & 0 & 0 & 0 & 0 & 0 & 0 \\
3 & 0 & -7 & 1 & 0 & 0 & 3 & 0 & 0 & 0 & 0 & 0 & 0 \\
0 & 3 & 2 & -10 & 0 & 0 & 0 & 3 & 0 & 0 & 0 & 0 & 2 \\
0 & 0 & 0 & 0 & -3 & 1 & 2 & 0 & 0 & 0 & 0 & 0 & 0 \\
0 & 1 & 0 & 0 & 2 & -7 & 0 & 2 & 0 & 2 & 0 & 0 & 0 \\
0 & 0 & 1 & 0 & 3 & 0 & -7 & 1 & 0 & 0 & 2 & 0 & 0 \\
0 & 0 & 0 & 1 & 0 & 3 & 2 & -8 & 0 & 0 & 0 & 2 & 0 \\
0 & 0 & 0 & 0 & 4 & 0 & 0 & 0 & -7 & 1 & 2 & 0 & 0 \\
0 & 0 & 0 & 0 & 0 & 4 & 0 & 0 & 2 & -9 & 0 & 2 & 0 \\
0 & 0 & 0 & 0 & 0 & 0 & 4 & 0 & 3 & 0 & -9 & 1 & 0 \\
0 & 0 & 0 & 0 & 0 & 0 & 0 & 4 & 0 & 3 & 2 & -10 & 1 \\
0 & 0 & 0 & 4 & 0 & 0 & 0 & 0 & 0 & 0 & 0 & 3 & -12
\end{pmatrix}$$

follows:

$$\begin{aligned}\tilde{\chi} = (&-16.033 - 0.125 - 13.455 - 3.169 - 3.418 - 11.71 \\ &- 11.238\ (-5.244 + 0.235\mathrm{i})\ (-5.244 - 0.235\mathrm{i}) \\ &- 6.683 - 7.739 - 8.55 - 9.393).\end{aligned}$$

Table 1.5 contains the matrix of eigenvectors $\tilde{Z}$.

Table 1.6 contains matrix (1.13) for $t = 0.3$. Further, matrix $R^* = (R^*(0), R^*(0.5), \ldots, R^*(3.0))$ is presented in Table 1.7. The latter presents reliability $R^*(t)$ with a step of 0.5 from 0 to 4. Matrix rows correspond to different initial states in accordance with matrix Ω.

Finally, we determine the optimal discipline for the service of the failed elements. Ten disciplines are compared. These are presented as columns in the following matrix:

$$\text{AllOrd} := \begin{pmatrix} 3 & 4 & 3 & 4 & 1 & 2 & 1 & 2 & 3 & 4 \\ 4 & 3 & 4 & 3 & 2 & 1 & 2 & 1 & 1 & 1 \\ 2 & 2 & 1 & 1 & 3 & 3 & 4 & 4 & 2 & 2 \\ 1 & 1 & 2 & 2 & 4 & 4 & 3 & 3 & 4 & 3 \end{pmatrix}.$$

Note that the above discipline $(1\ 2\ 3\ 4)^T$ has been considered (the fifth column of the last matrix). The results are presented in Table 1.8. It is assumed that initially, all the elements are working (state (1 1 1 1)). The rows correspond to the disciplines in the matrix AllOrd. The columns correspond to time t from 0 to 6 in steps of 0.5.

Table 1.5. Matrix of eigenvectors $\tilde{Z}$.

.036	−0.309	−0.011	−0.181	−0.503	−0.173	−0.146	0.231 − 0.01i	0.231 + 0.01i	0.245	−0.064	0.427	0.027
−0.14	−0.294	0.097	0.372	−0.031	0.093	0.371	0.274 + 0.068i	0.274 − 0.068i	−0.583	0.296	0.138	−0.253
0.081	−0.302	0.034	−0.176	−0.546	0.369	0.165	−0.31 − 0.102i	−0.31 + 0.102i	0.231	0.093	−0.632	0.056
0.439	−0.269	−0.456	0.357	−0.018	−0.294	−0.32	−0.302 + 0.012i	−0.302 − 0.012i	−0.543	−0.399	−0.366	0.105
−0.018	−0.305	−0.027	−0.177	−0.059	0.052	0.021	0.174 + 0.061i	0.174 − 0.061i	−0.016	−0.124	0.012	0.016
0.106	−0.289	0.103	0.357	0.31	0.165	−0.213	0.203 + 0.06i	0.203 − 0.06i	0.137	0.236	−0.112	0.113
0.063	−0.294	0.09	−0.163	−0.143	−0.308	0.019	−0.304 − 0.078i	−0.304 + 0.078i	−0.04	0.175	0.022	−0.107
−0.206	−0.285	−0.095	0.372	0.284	−0.093	0.134	−0.336	−0.336	0.049	−0.24	0.237	0.467
0.054	−0.29	0.125	−0.182	0.011	−0.177	0.059	0.275 − 0.073i	0.275 + 0.073i	−0.088	−0.031	−0.092	0.311
−0.186	−0.254	−0.258	0.302	0.347	−0.394	0.111	0.195 − 0.018i	0.195 + 0.018i	0.28	0.128	−0.231	−0.492
−0.114	−0.26	−0.22	−0.146	−0.036	0.511	−0.222	−0.196 − 0.144i	−0.196 + 0.144i	−0.123	0.194	0.162	−0.157
0.386	−0.261	0.244	0.349	0.336	0.381	0.243	−0.311 − 0.058i	−0.311 + 0.058i	0.138	−0.361	0.264	−0.441
−0.723	−0.157	0.751	0.28	0.109	−0.116	−0.724	−0.317 − 7.738i ×10⁻³	−0.317 + 7.738i ×10⁻³	−0.331	−0.629	−0.195	−0.346

Table 1.6. Matrix $R(0.3)$.

$$
\left(\begin{array}{cccccccc}
0.228 & 0.061 & 0.111 & 0.025 & 0.319 & 0.064 & 0.117 & 0.027 \\
0.115 & 0.189 & 0.053 & 0.073 & 0.145 & 0.176 & 0.054 & 0.076 \\
0.158 & 0.04 & 0.196 & 0.041 & 0.201 & 0.041 & 0.186 & 0.043 \\
0.07 & 0.11 & 0.083 & 0.112 & 0.088 & 0.109 & 0.083 & 0.122 \\
7.842 \times 10^{-3} & 0.012 & 0.022 & 6.224 \times 10^{-3} & 0.531 & 0.101 & 0.186 & 0.042 \\
0.018 & 0.057 & 0.014 & 0.024 & 0.221 & 0.253 & 0.081 & 0.113 \\
0.024 & 0.011 & 0.056 & 0.013 & 0.306 & 0.061 & 0.267 & 0.064 \\
0.015 & 0.037 & 0.027 & 0.041 & 0.14 & 0.169 & 0.129 & 0.192 \\
3.643 \times 10^{-3} & 6.948 \times 10^{-3} & 0.012 & 3.759 \times 10^{-3} & 0.372 & 0.072 & 0.133 & 0.03 \\
6.547 \times 10^{-3} & 0.027 & 7.03 \times 10^{-3} & 0.013 & 0.158 & 0.183 & 0.059 & 0.083 \\
8.363 \times 10^{-3} & 5.295 \times 10^{-3} & 0.026 & 7.261 \times 10^{-3} & 0.218 & 0.044 & 0.193 & 0.047 \\
7.055 \times 10^{-3} & 0.021 & 0.016 & 0.029 & 0.102 & 0.127 & 0.096 & 0.147 \\
0.02 & 0.043 & 0.031 & 0.07 & 0.034 & 0.055 & 0.04 & 0.081
\end{array}\right.
$$

$$
\left.\begin{array}{ccccc}
5.147 \times 10^{-3} & 0.01 & 0.018 & 6.042 \times 10^{-3} & 3.599 \times 10^{-3} \\
9.186 \times 10^{-3} & 0.041 & 0.011 & 0.022 & 0.014 \\
0.012 & 8.01 \times 10^{-3} & 0.04 & 0.012 & 7.818 \times 10^{-3} \\
9.263 \times 10^{-3} & 0.03 & 0.021 & 0.043 & 0.035 \\
0.013 & 0.021 & 0.037 & 0.011 & 1.255 \times 10^{-3} \\
0.028 & 0.088 & 0.023 & 0.041 & 6.029 \times 10^{-3} \\
0.037 & 0.017 & 0.087 & 0.023 & 3.228 \times 10^{-3} \\
0.024 & 0.061 & 0.045 & 0.073 & 0.013 \\
0.169 & 0.047 & 0.087 & 0.022 & 1.82 \times 10^{-3} \\
0.081 & 0.138 & 0.041 & 0.064 & 7.385 \times 10^{-3} \\
0.111 & 0.031 & 0.142 & 0.036 & 3.966 \times 10^{-3} \\
0.056 & 0.096 & 0.072 & 0.116 & 0.02 \\
0.014 & 0.033 & 0.024 & 0.061 & 0.049
\end{array}\right)
$$

Table 1.7. Matrix R^*.

$$
\begin{pmatrix}
1 & 0.982 & 0.932 & 0.878 & 0.825 & 0.775 & 0.727 & 0.683 & 0.642 & 0.603 & 0.566 & 0.532 \\
1 & 0.952 & 0.891 & 0.836 & 0.785 & 0.737 & 0.693 & 0.65 & 0.611 & 0.574 & 0.539 & 0.506 \\
1 & 0.966 & 0.913 & 0.859 & 0.807 & 0.758 & 0.712 & 0.669 & 0.628 & 0.59 & 0.554 & 0.521 \\
1 & 0.878 & 0.816 & 0.765 & 0.718 & 0.675 & 0.634 & 0.595 & 0.559 & 0.525 & 0.493 & 0.463 \\
1 & 0.975 & 0.922 & 0.867 & 0.815 & 0.765 & 0.719 & 0.675 & 0.634 & 0.596 & 0.559 & 0.525 \\
1 & 0.939 & 0.876 & 0.821 & 0.771 & 0.724 & 0.68 & 0.638 & 0.6 & 0.563 & 0.529 & 0.497 \\
1 & 0.947 & 0.89 & 0.837 & 0.786 & 0.738 & 0.693 & 0.651 & 0.612 & 0.575 & 0.54 & 0.507 \\
1 & 0.934 & 0.866 & 0.812 & 0.762 & 0.715 & 0.672 & 0.631 & 0.593 & 0.557 & 0.523 & 0.491 \\
1 & 0.934 & 0.878 & 0.825 & 0.775 & 0.728 & 0.684 & 0.642 & 0.603 & 0.567 & 0.532 & 0.5 \\
1 & 0.833 & 0.773 & 0.724 & 0.679 & 0.638 & 0.599 & 0.563 & 0.529 & 0.496 & 0.466 & 0.438 \\
1 & 0.842 & 0.788 & 0.74 & 0.695 & 0.653 & 0.613 & 0.576 & 0.541 & 0.508 & 0.477 & 0.448 \\
1 & 0.862 & 0.794 & 0.744 & 0.698 & 0.656 & 0.616 & 0.578 & 0.543 & 0.51 & 0.479 & 0.45 \\
1 & 0.518 & 0.476 & 0.446 & 0.418 & 0.393 & 0.369 & 0.347 & 0.325 & 0.306 & 0.287 & 0.27
\end{pmatrix}
$$

Table 1.8. Comparison of various disciplines.

$$
\mathrm{ROrd} := \begin{pmatrix}
1 & 0.887 & 0.719 & 0.577 & 0.462 & 0.37 & 0.297 & 0.238 & 0.19 & 0.152 & 0.122 & 0.098 \\
1 & 0.887 & 0.719 & 0.577 & 0.462 & 0.37 & 0.297 & 0.238 & 0.19 & 0.152 & 0.122 & 0.098 \\
1 & 0.887 & 0.717 & 0.574 & 0.459 & 0.367 & 0.293 & 0.234 & 0.187 & 0.15 & 0.12 & 0.096 \\
1 & 0.887 & 0.717 & 0.574 & 0.459 & 0.367 & 0.293 & 0.234 & 0.187 & 0.15 & 0.12 & 0.096 \\
1 & 0.934 & 0.866 & 0.812 & 0.762 & 0.715 & 0.672 & 0.631 & 0.593 & 0.557 & 0.523 & 0.491 \\
1 & 0.934 & 0.866 & 0.812 & 0.762 & 0.715 & 0.672 & 0.631 & 0.593 & 0.557 & 0.523 & 0.491 \\
1 & 0.933 & 0.863 & 0.807 & 0.755 & 0.708 & 0.663 & 0.622 & 0.583 & 0.546 & 0.512 & 0.48 \\
1 & 0.933 & 0.863 & 0.807 & 0.755 & 0.708 & 0.663 & 0.622 & 0.583 & 0.546 & 0.512 & 0.48 \\
1 & 0.926 & 0.844 & 0.777 & 0.718 & 0.663 & 0.613 & 0.566 & 0.523 & 0.483 & 0.447 & 0.413 \\
1 & 0.925 & 0.841 & 0.774 & 0.714 & 0.659 & 0.608 & 0.561 & 0.518 & 0.478 & 0.441 & 0.407
\end{pmatrix}
$$

We can see that discipline $(1\ 2\ 3\ 4)^T$ from the fifth row and $(2\ 1\ 3\ 4)^T$ from the sixth row yield the best results.

1.3 Matrix exponent

The examples considered used formula (1.6) to solve Eq. (1.1). There is another form of solution presentation using the *matrix exponent* [2,6,7].

Definition 1.1. The matrix exponent for a square matrix A is defined as

$$
\exp(A) = \sum_{n=0}^{\infty} \frac{1}{n!} A^n, \tag{1.17}
$$

where $A^0 = I$.

Property 1.1. The matrix exponent is finite for any finite matrix A.

Proof. Let c be the maximal absolute value of the elements of matrix A, k be a dimension of A, and E be a square matrix of dimension k from the units. Obviously, $A^n \leq (cE)^n$ for $n = 1, 2, \ldots$, we have

$$(cE)^2 = c^2 E^2 = c^2 kE,$$
$$(cE)^3 = cE(cE)^2 = cEc^2kE = ck(cE)^2 = c^3k^2E.$$

Therefore,

$$(cE)^n = c^n\, k^{n-1} E,$$
$$A^n \leq c^n\, k^{n-1} E = \frac{1}{k}(ck)^n E.$$

Let $n*$ be the first n, for which $q = \frac{(ck)^{n*}}{n*!} < 1$. Then for $n > n*$

$$\frac{1}{n!}A^n \leq \frac{1}{n!}\frac{1}{k}(ck)^n E = \frac{1}{n*!}(ck)^{n*}\frac{n*!}{n!}(ck)^{n-n*}\frac{1}{k}E \leq q\frac{1}{k}q^n E.$$

We have here a geometrical progression with the coefficient $q < 1$. It is known that its infinite sum converges. □

Property 1.2. The following equation takes place for the derivative with respect to the nonnegative variable t:

$$\frac{\partial}{\partial t}\exp(tA) = A\exp(tA). \tag{1.18}$$

Proof. We have

$$\frac{\partial}{\partial t}\exp(tA) = \frac{\partial}{\partial t}\sum_{n=0}^{\infty}\frac{1}{n!}(tA)^n = \sum_{n=1}^{\infty}\frac{1}{n!}A^n\frac{\partial}{\partial t}t^n = \sum_{n=1}^{\infty}\frac{1}{n!}A^n n t^{n-1}$$
$$= A\sum_{n=0}^{\infty}\frac{1}{n!}(tA)^n = A\exp(tA).$$

The property is proven. □

Let us consider an important relationship that expresses the matrix exponent (1.17) with eigenvalues $\chi_1, \chi_2, \ldots, \chi_k$ and eigenvectors $\beta_1, \beta_2, \ldots, \beta_k$ of the matrix A. As before, let B be the matrix with columns $\beta_1, \beta_2, \ldots, \beta_k$. Further, we assume that all the eigenvectors are different. In this case, the inverse matrix B^{-1} exists, and from (1.2), we have a spectral expansion of matrix A:

$$A = B\text{diag}(\chi)B^{-1}.$$

Therefore,

$$A^n = B\text{diag}(\chi)B^{-1} \cdots B\text{diag}(\chi)B^{-1} = B\text{diag}(\chi)^n B^{-1},$$

$$\begin{aligned}\exp(A) &= \sum_{n=0}^{\infty} \frac{1}{n!} A^n = \sum_{n=0}^{\infty} \frac{1}{n!} B\text{diag}(\chi)^n B^{-1} \\ &= B \sum_{n=0}^{\infty} \frac{1}{n!} \begin{pmatrix} \chi_1^n & 0 & \ldots & 0 & 0 \\ 0 & \chi_2^n & & 0 & 0 \\ & \vdots & \ddots & \vdots & \\ 0 & 0 & \ldots & \chi_{k-1}^n & 0 \\ 0 & 0 & & 0 & \chi_k^n \end{pmatrix} B^{-1}. \\ &= B \begin{pmatrix} \exp(\chi_1) & 0 & \ldots & 0 & 0 \\ 0 & \exp(\chi_2) & & 0 & 0 \\ & \vdots & \ddots & \vdots & \\ 0 & 0 & \ldots & \exp(\chi_{k-1}) & 0 \\ 0 & 0 & & 0 & \exp(\chi_k) \end{pmatrix} B^{-1}.\end{aligned}$$

Finally, if $\text{diag}(\exp(\chi)) = \text{diag}(\exp(\chi_1), \exp(\chi_2), \ldots, \exp(\chi_k))$ then

$$\exp(A) = B\text{diag}(\exp(\chi))B^{-1}. \tag{1.19}$$

Therefore, if $A = \lambda - \text{diag}(\Lambda)$ is the *generator* of the Markov chain, then instead of the formula (1.9) we have the alternative presentation

$$P(t) = B\text{diag}(\exp(\chi t))B^{-1} = \exp(tA), \quad t \geq 0. \tag{1.20}$$

Example 1.4 (Poisson process). In conclusion, we consider the example of a continuous-time Markov chain with an infinite number of states. Here, the Poisson process was utilized. It is the continuous-time Markov chain $J(t)$, having states $0, 1, \ldots$; jumps are possible from state i to state $i+1$ only, $i = 0, 1, \ldots$; the same intensity λ of the transitions.

We show how the main formula for the Poisson process

$$P\{J(t) = \eta \mid J(0) = 0\} = \frac{1}{\eta!}(\lambda t)^{\eta}\exp(-\lambda t), \quad \eta = 0, 1, \ldots, \quad t \geq 0. \tag{1.21}$$

can be given, using the matrix exponent considered above.

Let us now consider a restricted number of states. In this case, the generator is expressed as

$$A = \begin{pmatrix} -\lambda & \lambda & 0 & & 0 & 0 & 0 \\ 0 & -\lambda & \lambda & \cdots & 0 & 0 & 0 \\ 0 & 0 & -\lambda & & 0 & 0 & 0 \\ & \vdots & & \ddots & & \vdots & \\ 0 & 0 & 0 & & -\lambda & \lambda & 0 \\ 0 & 0 & 0 & \cdots & 0 & -\lambda & \lambda \\ 0 & 0 & 0 & & 0 & 0 & -\lambda \end{pmatrix}.$$

Only the zero rows of the matrix product A^n are of interest to us. Let us denote one by $a_0, a_1, \ldots$ for $n = 0, 1, \ldots$. We have the following sequence:

$$\begin{aligned} a_0 &= (1 \quad 0 \quad 0 \quad 0 \quad 0 \quad 0 \quad 0 \quad 0 \quad \ldots), \\ a_1 &= (-\lambda \quad \lambda \quad 0 \quad 0 \quad 0 \quad 0 \quad 0 \quad 0 \quad \ldots), \\ a_2 &= (\lambda^2 \quad -2\lambda^2 \quad \lambda^2 \quad 0 \quad 0 \quad 0 \quad \ldots), \\ a_3 &= (-\lambda^3 \quad 3\lambda^3 \quad -3\lambda^3 \quad \lambda^3 \quad 0 \quad 0 \quad \ldots), \\ a_4 &= (\lambda^4 \quad -4\lambda^4 \quad 6\lambda^4 \quad -4\lambda^4 \quad \lambda^4 \quad 0 \quad \ldots). \end{aligned}$$

If $A_{\langle\eta\rangle}$ is the ηth row of matrix A, then the general formulas are as follows:

$$a_0 = (1 \quad 0 \quad 0 \quad 0 \quad \ldots),$$

$$a_{n+1} = a_n\, A = \sum_{\eta=0}^{n} a_{n,\eta}\, A_{\langle\eta\rangle}, \quad n = 0, 1, \ldots$$

We can see that the components of the current row are calculated using those of the previous row. Specifically, the ηth component, $\eta = 0, 1, \ldots$, of the nth row, $n > 0$, equals the sum of the ηth component of the previous row multiplied by $-\lambda$ and the $(\eta - 1)$th component of the previous row multiplied by λ:

$$a_{n+1,\eta} = -\lambda\, a_{n,\eta} + \lambda\, a_{n,\eta-1} = \lambda(a_{n,\eta-1} - a_{n,\eta}),$$
$$n = 0, 1, \ldots, \eta = n, n+1, \ldots$$

Lemma 1.1. *The following relation takes place:*

$$a_{n,\eta} = \begin{cases} 0, & \eta > n, \\ (-1)^{n+\eta} \dfrac{n!}{\eta!(n-\eta)!} \lambda^n, & \text{otherwise.} \end{cases}$$

Proof. The relationship is true for $n = 0$ and 1. Using mathematical induction, we have

$$a_{n+1,\eta} = \lambda(a_{n,\eta-1} - a_{n,\eta}) = \lambda\left((-1)^{n+1+\eta} \frac{n!}{(\eta-1)!(n-(\eta-1))!} \lambda^n - (-1)^{n+\eta} \frac{n!}{\eta!(n-\eta)!} \lambda^n\right)$$

$$= (-1)^{n+1+\eta} \lambda^{n+1} n! \frac{\eta - (-1)(n+1-\eta)}{\eta!(n+1-\eta)!}$$

$$= (-1)^{n+1+\eta} \lambda^{n+1} n! \frac{n+1}{\eta!(n+1-\eta)!}. \ \#$$

For $\tilde{a}_{n,\eta} \doteq \frac{1}{n!} a_{n,\eta} t^n, n = 0, 1, \ldots, \eta = n,\ n = 1, \ldots$, we have

$$\tilde{a}_{n,\eta} = \begin{cases} 0, & \eta > n, \\ (-1)^{n+\eta} \dfrac{1}{\eta!(n-\eta)!} (\lambda t)^n, & \text{otherwise.} \end{cases}$$ □

Lemma 1.2. *The relation* (1.21) *takes place.*

Proof.

$$P\{X(t) = \eta \,|\, X(0) = 0\} = (e^{tA})_{0,\eta} = \sum_{n=0}^{\infty} \tilde{a}_{n,\eta}$$

$$= \sum_{n=\eta}^{\infty} (-1)^{n+\eta} \frac{1}{\eta!(n-\eta)!} (\lambda t)^n$$

$$= \frac{1}{\eta!} (\lambda t)^\eta \sum_{n=\eta}^{\infty} (-\lambda t)^{n-\eta} \frac{1}{(n-\eta)!} = \frac{1}{\eta!} (\lambda t)^\eta \exp(-\lambda t).\ \#$$

This result can be obtained more simply if the generation function $\varphi(z,t) = E(z^{N(t)}) = \sum_{\eta=0}^{\infty} z^\eta P\{N(t) = \eta \,|\, N(0) = 0\}$ is used. For time $t + \Delta t$, the value $\varphi(z,t)$ does not vary with probability $1 - \lambda\Delta t + o(\Delta t)$ and will be equal $z\varphi(z,t)$ with probability $\lambda\Delta t + o(\Delta t)$. Therefore,

$$\varphi(z, t+\Delta t) = \varphi(z,t)(1 - \lambda\Delta t) + z\varphi(z,t)\lambda\Delta t + o(\Delta t).$$

Further

$$\frac{\partial}{\partial t}\varphi(z,t) = -\lambda\varphi(z,t) + z\varphi(z,t)\lambda = \varphi(z,t)\lambda(z-1),$$

$$\frac{1}{\varphi(z,t)} \frac{\partial}{\partial t}\varphi(z,t) = \lambda(z-1),$$

$$\frac{\partial}{\partial t}\ln(\varphi(z,t)) = \lambda(z-1),$$

$$\ln(\varphi(z,t)) = \lambda t(z-1),$$

$$\varphi(z,t) = e^{\lambda t(z-1)} = \exp(-\lambda t) \sum_{\eta=0}^{\infty} \frac{1}{\eta!} (\lambda t)^\eta z^\eta.\ \#$$

□

1.4 Phase-type distribution

This distribution was introduced by Neuts in 1981 [2]. It can be interpreted as follows. Continuous-time Markov chain with $k + 1$ states is considered. The last $(k+1)$th state is absorbing one: if the chain gets to this state, that one remains in the same state forever.

It is supposed that the absorbing state is reached from each other state with the probability 1.

The matrix of transfer intensities between states is as follows:

$$\tilde{\lambda} = \begin{pmatrix} \lambda & r \\ \mathbf{0} & \xi \end{pmatrix},$$

where λ is $k \times k$ a matrix of intensities, $r = (\lambda_{1,k+1} \quad \cdots \quad \lambda_{k,k+1})^T$ is a vector of intensities of transfers to the last state, $\mathbf{0}$ is the zero vector of the dimension k, ξ is some positive number.

Let $\mathbf{1}$ be the unit-vector of dimension k and

$$\Lambda = \lambda \mathbf{1}.$$

The matrix

$$G = \lambda - (\Lambda + \text{diag}(r))$$

has non-negative off-diagonal elements, non-positive diagonal elements, and at least one negative diagonal element. Such a generator is called a *sub-generator.*

Let α be a probability vector, which defines the initial state of the Markov chain. *The phase-type density function $f(t)$ is the first-passage-time density to state $k+1$ with the initial distribution α* [2]. It is follows from (1.20) that

$$f(t) = \alpha^T \exp(tG) r, \quad t \geq 0. \tag{1.22}$$

It is a classical definition of the phase-type density function, which uses the matrix exponent.

Another definition uses spectral characteristics of the generator G, namely a vector of eigenvalues χ and a matrix of eigenvectors B. In this case, the following presentation takes place with respect to (1.19) and (1.20):

$$f(t) = \alpha^T B \text{diag}(\exp(\chi t)) B^{-1} r, \quad t \geq 0. \tag{1.23}$$

The last presentation has an obvious advantage: it is necessary to calculate the vector χ and the matrices B and B^{-1} only once. The alternate presentation requests a calculation of the matrix exponent for each value of time t.

If the vector $\chi = (\chi_1 \quad \cdots \quad \chi_k)^T$ is strictly negative ($\chi_i < 0$, $i = 1, \ldots, k$), then all moments μ_ν, $\nu = 1, 2, \ldots$, of the phase-type distribution exist and are given by the formula

$$\mu_\nu = \nu! \alpha^T \sum_{i=1}^{k} |\chi_i|^{-(\nu+1)} \beta_i \tilde{\beta}_i r. \tag{1.24}$$

Actually

$$\begin{aligned}
\mu_\nu &= \int_0^\infty t^\nu f(t) dt = \int_0^\infty t^\nu \sum_{i=1}^{k} \alpha^T \exp(\chi_i t) \beta_i \tilde{\beta}_i r \, dt \\
&= \alpha^T \sum_{i=1}^{k} |\chi_i|^{-1} \int_0^\infty t^\nu |\chi_i| \exp(-|\chi_i| t) \, dt \beta_i \tilde{\beta}_i r \\
&= \alpha^T \sum_{i=1}^{k} |\chi_i|^{-1} \, |\chi_i|^{-\nu} \nu! \beta_i \tilde{\beta}_i r.
\end{aligned}$$

The formula (1.24) can be presented by the matrix exponent. Because $\tilde{\beta}_i \, \beta_j = 1 \; if \; i = j$ and $\tilde{\beta}_i \, \beta_j = 0$ otherwise; then,

$$\begin{aligned}
\mu_\nu &= \sum_{i=1}^{k} \alpha^T |\chi_i|^{-1} |\chi_i|^{-\nu} \nu! \beta_i \tilde{\beta}_i r = \nu! \alpha^T \sum_{i=1}^{k} (-\chi_i)^{-(\nu+1)} (\beta_i \tilde{\beta}_i)^{\nu+1} r \\
&= \nu! \alpha^T \sum_{i=1}^{k} (\beta_i (-\chi_i^{-1}) \tilde{\beta}_i)^{\nu+1} r = \nu! \alpha^T \left(\sum_{i=1}^{k} (\beta_i (-\chi_i^{-1}) \tilde{\beta}_i) \right)^{\nu+1} r \\
&= \nu! \alpha^T \left(B \, (\mathrm{diag}(-\chi))^{-1} \tilde{B}\right)^{\nu+1} r = \nu! \alpha^T \left(-B \, (\mathrm{diag}(\chi))^{-1} \tilde{B}\right)^{\nu+1} r \\
&= \nu! \alpha^T \left(-G^{-1}\right)^{\nu+1} r.
\end{aligned}$$

Example 1.5. Let us consider the following data:

$$\alpha = \begin{pmatrix} 0.5 \\ 0.3 \\ 0.2 \\ 0.0 \end{pmatrix}, \quad \lambda = \begin{pmatrix} 0 & 0.1 & 0.2 & 0 \\ 0.3 & 0 & 0.1 & 0 \\ 0.2 & 0 & 0 & 0.1 \\ 0 & 0.1 & 0.2 & 0 \end{pmatrix}, \quad r = \begin{pmatrix} 0 \\ 0.2 \\ 0.3 \\ 0.5 \end{pmatrix}.$$

The following generator takes place for the presented data:

$$G = \begin{pmatrix} -0.3 & 0.1 & 0.2 & 0 \\ 0.3 & -0.6 & 0.1 & 0 \\ 0.2 & 0 & -0.6 & 0.1 \\ 0 & 0.1 & 0.2 & -0.8 \end{pmatrix}.$$

Now we can use formula (1.22).

The spectral characteristics of the generator G are as follows:

$$\chi^T = (-0.13 \quad -0.89 \quad -0.669 \quad -0.611),$$

$$B = \begin{pmatrix} 0.737 & -0.138 & -0.146 & -0.187 \\ 0.545 & 2.793 \times 10^{-3} & 0.883 & -0.686 \\ 0.353 & 0.406 & -0.173 & 0.633 \\ 0.187 & -0.903 & 0.411 & 0.306 \end{pmatrix},$$

$$B^{-1} = \begin{pmatrix} 0.948 & 0.217 & 0.48 & 0.072 \\ -0.431 & 0.403 & 0.668 & -0.74 \\ -0.99 & 0.89 & 0.491 & 0.375 \\ -0.523 & -0.137 & 1.017 & 0.537 \end{pmatrix}.$$

Now we can use formula (1.23).

Figure 1.3 contains the graph of the density $f(t)$, which is calculated by formula (1.23), and analogous graphs calculated by formula (1.22). The last two graphs are denoted $fM(t,\ n\max)$, where $n\max$ is an upper limit of the sum in (1.17). We see that the graphs of densities $f(t)$, and f M$(t$, 100) coincide fully, but 15 is an insufficient number of addends. This fact illustrates an advantage of the spectral presentation (1.23).

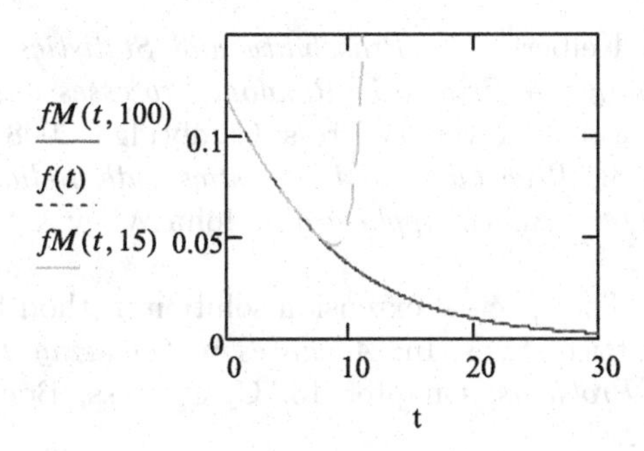

Fig. 1.3. Graphs of phase-type densities functions.

Calculations by formula (1.24) give the following values of the expectation, the second moment, and the standard deviation of the distribution:

$$\mu_1 = 7.884, \quad \mu_2 = 122.083, \quad \sigma = \sqrt{\mu_2 - \mu_1^2} = 7.742.$$

It is interesting to remark $\mu_0 = 1$, thus, the norming's condition of the density is fulfilled.

References

[1] Bellman, R., *Introduction to Matrix Analysis*, McGraw-Hill Book Company, New York, Toronto, London, 1960.

[2] Neuts, M.F., *Matrix-Geometric Solutions in Stochastic Models — An Algorithmic Approach.* Johns Hopkins University Press, Baltimore and London, 1981.

[3] Anderson, W.J., *Continuous-time Markov Chains: An Applications-Oriented Approach.* Springer-Verlag, New York, Berlin, Heidelberg, London, Paris, Tokyo, Hong Kong, Barcelona, 1991.

[4] Bharucha-Reid, A.T., *Elements of the Theory of Markov Processes and their Applications.* McGraw-Hill, New York, 1960.

[5] Chen, M.-F., Mao, Y.-H., *Introduction to Stochastic Processes.* World Scientific, New Jersey, 2021.

[6] Kijima, M., *Markov Processes for Stochastic Modeling,* The University Press, Cambridge, UK, 1997.

[7] Pacheco, A., Tang, L.C., and Prabhu, N.U., *Markov-Modulated Processes & Semiregenerative Phenomena,* World Scientific, New Jersey, London, 2009.

[8] Stroock, D.W., *An Introduction to Markov Processes.* Springer, Berlin, 2005.

[9] Suhov, Y., Kelbert, M., *Probability and Statistics by Example: II Markov Chains: A Primer in Random Processes and their Applications.* Cambridge University Press, Cambridge, 2008.

[10] Triwedi, K.S., *Probability and Statistics with Reliability, Queuing, and Computer Science Applications,* John Wiley & Sons, New York, 2001.

[11] Mitrani, I., The spectral extension solution method for Markov processes on lattice strips. In: *Advances in Queueing Theory, Methods and Open Problems,* Chapter 13, CRC Press, Boca Raton, 1995, pp. 337–352.

[12] Pontryagin, L.S., *Ordinary Differential Equations,* Nauka, Moscow, 1963 (in Russian).

Chapter 2

Markov Modulated Markov Chains

2.1 Definition

When probabilistic models of an object's operation are elaborated, it is often necessary to take into account the influence of exterior factors, the *random environment*. In this case, the process is described by two components: the state of the external environment and the state of the object properly. We denote the first component by $J(t)$ and call it the *Markovian component* and denote the second component by $X(t)$ and call it the *main component*. Both components form a process: $Y(t) = (J(t), X(t))$. If an external environment is described by a continuous-time Markov chain, it suggests *Markov-modulated processes*[1–8]. In this chapter, we consider a case in which both components are continuous-time Markov chains.

Let us set the transition intensities between states of a Markov chain $J(t)$ as $\lambda_{i,j}$ which describes the external environment, n denotes its states number, and a matrix of transition intensities $\lambda = (\lambda_{i,j})_{n \times n}$. Transitions between states do not depend on the state of the Markov chain $X(t)$, which describes the object properly. A state number of chain's $X(t)$ equals m. Transition intensities of this chain depend on the state of chain $J(t)$. If the last chain has state i, then a matrix of transition intensities for chain $X(t)$ will be denoted $\mu(i) = (\mu(i)_{r,s})_{m \times m}$.

As before, the problem consists of determining the transition probabilities between the states of the composite chain $Y(t) =$

$(J(t), X(t))$. The number of states in this chain is equal to nm. It is clear that $Y(t)$ is a finite continuous-time Markov chain. Let us find the transient intensities between states (i, r) and (j, s) using the notation $\nu_{(i,r),(j,s)}$.

In accordance with our suppositions, the following relations for $i, j = 1, \ldots, n$; $r, s = 0, \ldots, m-1$ occur:

$$\nu_{(i,r),(i,s)} = \mu(i)_{r,s}, \quad r \neq s; \quad \nu_{(i,r),(j,r)} = \lambda_{i,j}, \quad i \neq j; \quad \nu_{(i,r),(i,r)} = 0. \tag{2.1}$$

The corresponding matrix is on the order $nm \times nm$. It is a block-matrix A with blocks $A(r, s)$, r, $s = 0, \ldots, m-1$:

$$A = \begin{pmatrix} A(0,0) & A(0,1) & \cdots & A(0,m-2) & A(0,m-1) \\ A(1,0) & A(1,1) & \cdots & A(1,m-2) & A(1,m-1) \\ \cdots & \cdots & \cdots & \cdots & \cdots \\ A(m-2,0) & A(m-2,1) & \cdots & A(m-2,m-2) & A(m-2,m-1) \\ A(m-1,0) & A(m-1,1) & \cdots & A(m-1,m-2) & A(m-1,m-1) \end{pmatrix}.$$

Each block is a square matrix of the order n. Diagonal matrices $A(r, r)$, $r = 0, \ldots, m-1$ coincide with the matrix $\lambda = (\lambda_{i,j})_{n \times n}$ of transition intensities for the chain $J(t)$. Non-diagonal blocks $A(r, s)$, $r \neq s$, are diagonal matrices with the values $\mu(i)_{r,s}$, $i = 1, \ldots n$ on the diagonal:

$$A(r,s) = \begin{pmatrix} \mu(1)_{r,s} & 0 & \cdots & 0 \\ 0 & \mu(2)_{r,s} & \cdots & 0 \\ \cdots & \cdots & \cdots & \cdots \\ 0 & 0 & \cdots & \mu(n)_{r,s} \end{pmatrix}.$$

Obviously, other presentations of the matrices are possible, for example, using blocks $A(i, j)$, $i, j = 1, \ldots, n$ of the dimension $m \times m$.

Now we use unary enumeration of the states, instead of the binary enumeration. We wish to know which ordered unary number *Num*(i, r) has the pair (i, r). If the argument i is modified more often than argument r, then

$$Num(i, r) = rn + i, \quad i \in \{1, \ldots, n\}, \quad r \in \{0, \ldots, m-1\}. \tag{2.2}$$

An inverse transition from $Num(i,r)$ to i and r is the following:

$$i(Num) = \mathrm{mod}(Num, n), r = (Num - i(Num))/n, \tag{2.3}$$

where $\mathrm{mod}(N(i,r), n)$ is the remainder on dividing Num by n.

Obviously, other presentations of the matrices are possible, for example, using blocks $A(i,j)$, $i,j = 1,\ldots,n$ of the dimension $m \times m$.

Example 2.1 (Machine repair problem faced when considering the existence of a random external environment). Let us consider Example 1.2. We assume that a random environment exists that changes the intensity of machine failure. This external environment is described by a Markov chain $J(t)$ with three states ($n = 3$) and the following matrix of the transient intensities between states:

$$\lambda = \begin{pmatrix} 0 & 1 & 2 \\ 2 & 0 & 3 \\ 3 & 5 & 0 \end{pmatrix}.$$

As before, there are two workers. To reduce the dimensions of the matrices used, we assume that the number of machines is three (instead of the previous four). Therefore, the number of failed machines $X(t)$ can range from 0 to 3, so the number of states $m = 4$.

The state of the external environment does not affect the repair intensity, one equals 2, as before. The intensity of machine failure depends on the state of the external environment. Specifically, it equals 0.5 for the first state, 1 for the second state, and 1.5 for the third state.

Now we have the following expressions for the matrices of repair intensities for different states of the random environment:

$$\mu(1) = \begin{pmatrix} 0 & 1.5 & 0 & 0 \\ 2 & 0 & 1 & 0 \\ 0 & 4 & 0 & 0.5 \\ 0 & 0 & 4 & 0 \end{pmatrix},$$

$$\mu(2) = \begin{pmatrix} 0 & 3 & 0 & 0 \\ 2 & 0 & 2 & 0 \\ 0 & 4 & 0 & 1 \\ 0 & 0 & 4 & 0 \end{pmatrix},$$

$$\mu(3) = \begin{pmatrix} 0 & 4.5 & 0 & 0 \\ 2 & 0 & 3 & 0 \\ 0 & 4 & 0 & 1.5 \\ 0 & 0 & 4 & 0 \end{pmatrix}.$$

These matrices allow for the calculation of the matrix of transient intensities (2.1) of the Markov chain and the corresponding generator. It allows the calculation of the matrix $P(t)$ of the non-stationary probabilities of the states using formula (1.9). We restrict ourselves to consideration of dependence for an average number of the failed machines $E(X(t))$ at time t, on an initial state of the environment *i0*, and on an initial number of the failed machines $j0$. Obviously,

$$E(X(t)|J(0) = i0, X(0) = j0) = \sum_{i=1}^{n} \sum_{j=0}^{m-1} jP_{(i0,j0),(i,j)}. \tag{2.4}$$

This dependence is presented in Table 2.1 if all the machines were operated initially ($j0 = 0$).

Table 2.1. Average number of failed machines $E(X(t))$ as a function of time t and an initial state of the external environment $i0$.

t	0.2	0.4	0.6	0.8	1.0	1.2	1.4	1.6
$i0 = 1$	0.90	0.89	0.902	0.916	0.926	0.933	0.938	0.941
$i0 = 2$	1.498	1/246	1.114	1.043	1/002	0.979	0.965	0.957
$i0 = 3$	2.282	1.777	1.454	1.254	1.132	1.058	1.013	

t	1.8	2.0	2.2	2.4	2.6	2.8	3.0	3.2
$i0 = 1$	0.943	0.944	0.945	0.945	0.945	0.945	0.945	0.945
$i0 = 2$	0.953	0.950	0.948	0.947	0.946	0.946	0.946	0.946
$i0 = 3$	0.97	0.96	0.954	0.951	0.949	0.947	0.947	0.946

2.2 Markov-modulated Poisson process

The Markov-modulated Poisson process has been examined by many authors [1,4,6,7]. It is supposed that the Poisson process (see Example 1.4) $X(t)$ operates in an external random environment

$J(t)$, having k states and intensity $\lambda_{i,j}$ of a transition from state i to state j. Process $J(t)$ is the continuous-time Markov chain described at the beginning of the previous chapter. As described in Section 1.2, we set $\Lambda_i = \sum_{j=1}^{k} \lambda_{i,j}$.

The intensity of the Poisson process $X(t)$ depends on the state of the external environment $J(t)$. This equals α_i for the ith state. We are interested in the distribution of $X(t)$ and $J(t)$ at time t if the ith state takes place initially. The corresponding probabilities are expressed as $P\{X(t) = \eta, J(t) = j | X(0) = 0, J(0) = i\}$.

We use the generation function for the probabilities. Let $\delta_j(t)$ be the indicator of a random event $\{J(t) = j\}$

$$\delta_j(t) = \begin{cases} 1, & \text{if event } \{J(t) = j\} \text{ occurs,} \\ 0, & \text{if not.} \end{cases}$$

Now the generation function is determined as

$$\begin{aligned} \psi_{i,j}(z,t) &= E(z^{X(t)}\delta_j(t)|J(0) = i) \\ &= \sum_{\eta=0}^{\infty} z^{\eta} P\{X(t) = \eta, J(t) = j|X(0) = 0, J(0) = i\}, z, \quad t \geq 0. \end{aligned}$$

For the value $\psi_{i,j}(z, t + \Delta t)$, considering the time increment Δt: changes are absent with the probability $1 - (\Lambda_i + \alpha_i)\Delta t + o(\Delta t)$; a new state of the Poisson process $X(t)$ takes place with the probability $\alpha_i \Delta t + o(\Delta t)$; the state i of the external random environment $J(t)$ is replaced by the state η with the probability $\lambda_{i,\eta}\Delta t + o(\Delta t)$. Therefore,

$$\begin{aligned} \psi_{i,j}(z, t + \Delta t) &= (1 - (\Lambda_i + \alpha_i)\Delta t)\psi_{i,j}(z,t) + \alpha_i \Delta t z \psi_{i,j}(z,t) \\ &\quad + \sum_{\eta=1}^{k} \lambda_{i,\eta}\, \Delta t \psi_{\eta,j}(z,t) + \mathrm{o}(\Delta \mathrm{t}). \end{aligned} \tag{2.5}$$

Now we have the following for the generation function $\psi_{i,j}(z,t)$ and the corresponding matrix $\Psi(z,t) = (\psi_{i,j}(z,t))$:

$$\frac{\partial}{\partial t}\psi_{i,j}(z,t) = -(\Lambda_i + \alpha_i)\psi_{i,j}(z,t) + \alpha_i z \psi_{i,j}(z,t) + \sum_{\eta=1}^{k} \lambda_{i,\eta}\, \psi_{\eta,j}(z,t).$$

$$\frac{\partial}{\partial t}\Psi(z,t) = -\mathrm{diag}(\Lambda + \alpha)\Psi(z,t) + \mathrm{diag}(\alpha) z \Psi(z,t) + \lambda\Psi(z,t).$$

$$\frac{\partial}{\partial t}\Psi(z,t) = (\lambda - \mathrm{diag}(\alpha(1 - z) - \Lambda))\Psi(z,t).$$

A solution is given by a matrix exponent (see (1.20) and (1.17)):

$$\Psi(z,t) = \exp(t(\lambda - \mathrm{diag}(\alpha(1-z) - \Lambda)))$$
$$= \sum_{n=0}^{\infty} \frac{1}{n!}(t(\lambda - \mathrm{diag}(\alpha(1-z) - \Lambda)))^n. \quad (2.6)$$

Note that if $z = 1$, this formula provides the matrix of transition probabilities (1.20) for process $J(t)$.

We calculate the average of the Poisson process $X(t)$ jointly with the probabilities of $J(t)$ states:

$$(E(X(t))_{k\times k} = \frac{\partial}{\partial z}\Psi(z,t)\Big|_{z=1}$$
$$= \frac{\partial}{\partial z}\exp\left(t(\lambda - \mathrm{diag}(\alpha(1-z) + \Lambda))\right)\Big|_{z=1}$$
$$= \frac{\partial}{\partial z}\sum_{n=0}^{\infty}\frac{1}{n!}\left(t\left(\lambda - \mathrm{diag}(\alpha(1-z) + \Lambda)\right)\right)^n\Big|_{z=1}$$
$$= \frac{\partial}{\partial z}\sum_{n=0}^{\infty}\frac{1}{n!}\left(t\left(\lambda - \mathrm{diag}(\Lambda) + \mathrm{diag}(\alpha(z-1))\right)\right)^n\Big|_{z=1}$$
$$= t\mathrm{diag}(\alpha) + \frac{\partial}{\partial z}\sum_{n=2}^{\infty}\frac{1}{n!}t^n\sum_{\eta=0}^{n}\frac{n!}{\eta!(n-\eta)!}(\lambda - \mathrm{diag}(\Lambda))^\eta$$
$$\times\ (\mathrm{diag}(\alpha(z-1)))^{n-\eta}\Big|_{z=1}$$
$$= t\mathrm{diag}(\alpha) + \sum_{n=2}^{\infty}\frac{1}{n!}t^n\sum_{\eta=0}^{n-1}\frac{n!}{\eta!(n-\eta)!}(\lambda - \mathrm{diag}(\Lambda))^\eta(n-\eta)$$
$$\times\ \mathrm{diag}(\alpha)(\mathrm{diag}(\alpha(z-1)))^{n-\eta-1}\Big|_{z=1}$$
$$= t\mathrm{diag}(\alpha) + \sum_{n=2}^{\infty}\frac{1}{(n-1)!}t^n(\lambda - \mathrm{diag}(\Lambda))^{n-1}\mathrm{diag}(\alpha).$$

The matrix $E(X(t)$ of elements $E\big((X(t)\delta_j(t)|J(0) = i\big)$ is the following:

$$E(X(t) = t\exp\Big(t(\lambda - \text{diag}(\Lambda))\Big)\,\text{diag}(\alpha). \tag{2.7}$$

If the final state j is not of interest then we have a column-vector

$$E^*(X(t)) = E(X(t))\Delta, \tag{2.8}$$

where Δ is the vector-column from units of the dimension k.

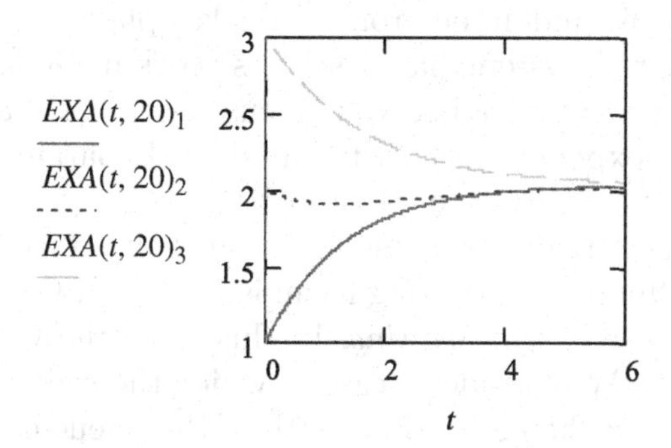

Fig. 2.1. Poisson process-specific average.

It is often convenient to consider a specific average related to the unit of time. We obtain these values by dividing the corresponding equation by t.

Example 2.2. The matrix of transition intensities for the Markov chain $J(t)$ is as follows:

$$\lambda = \begin{pmatrix} 0 & 0.2 & 0.3 \\ 0.4 & 0 & 0.2 \\ 0.2 & 0.2 & 0 \end{pmatrix}.$$

The intensity of the Poisson process $X(t)$ for different states of the Markov chain $J(t)$ is given by the vector $\alpha = (\alpha_1, \alpha_2, \alpha_3) = (1, 2, 3)$.

Figure 2.1 contains the specific averages $EXA(t, nmax)_i = E^*(X(t))_i/t$ for these data. Here, i is the initial state of the Markov chain $J(t)$, t is time, $n$max is the number of addends in the infinite sum (2.7). We observe how various trajectories tend to be stationary.

2.3 Markov-modulated Erlang systems

Erlang systems are the first stochastic models considered in queueing theory [8–10]. We described Erlang systems for the case of random environment existence.

The external random environment $J(t)$ is presented as the continuous-time irreducible Markov chain with k states. The matrix of transient intensities between states is $\lambda = (\lambda_{i,j.})$, $i, j = 0, \ldots, k-1$.

Customers arrive according to a Poisson process with intensity α_i if the ith state of random environment takes place.

The considered system has s servers working independently of each other. Each server can serve one customer simultaneously. Service times are exponentially distributed with parameter μ_i if the random environment has the ith state.

Two cases take place with respect to the waiting room. It is *a system with rejections* if the waiting room is absent. If the waiting room exists, it is *a system with waiting.* In this case, the waiting room is called *the queue.* We consider a case in which the waiting room has n places for waiting; therefore, the length of the queue cannot be more than n.

Let $X(t)$ be the number of customers in the system at time t. Hence $Y(t) = (J(t), X(t))$ is a two-dimensional continuous-time finite Markov chain. We wish to calculate the probabilities $P\{J(t) = j, X(t) = x|J(0) = i, X(0) = 0\}$, where $i \in \{0, \ldots, k-1\}$, $x \in \{0, \ldots, s+n\}$.

Usual reasoning gives the following expressions for non-zero elements of the generator G:

$$\begin{aligned}
G_{(i.x),(j,x)} &= \lambda_{i,j}, j \neq i; \\
G_{(i.x),(i.x+1)} &= \alpha_i, x = 0, \ldots, s+n-1; \\
G_{((i.x)),(i.x-1)} &= \mu_i x, x = 1, \ldots, s-1; \\
G_{((i.x)),(i.x-1)} &= \mu_i s, x = s, \ldots, s+n; \\
G_{(i.x),(i,x)} &= -(\Lambda_i + \mu_i x + \alpha_i), x = 0, \ldots, s-1; \\
G_{(i.x),(i,x)} &= -(\Lambda_i + \mu_i s + \alpha_i), x = s, \ldots, s+n,
\end{aligned} \tag{2.9}$$

where $\Lambda_i = \lambda_{i,0} + \cdots + \lambda_{i,k-1}$.

The total number of states equals $k(s+n+1)$. Let us enumerate ones using the formula

$$Num(i,x) = xk + i, i \in \{0,\ldots,k-1\}, x \in \{0,\ldots,s+n\}.$$

The state (i,x) is renewed by its order number *Num* as follows:

$$i(Num) = \mathrm{mod}(Num, k),$$
$$x(Num) = (Num - i(Num))/k,$$

where $\mathrm{mod}(Num, k)$ is the remainder from dividing *Num* by k.

Now we can present the generator G in the form of square matrix with dimension $k(s+n+1)$. It allows the calculation of probabilities $P\{J(t) = j, X(t) = x | J(0) = i, X(0) = 0\}$ using spectral characteristics of the generator.

Let us consider stationary probabilities $p_{x,i}, i \in \{0,\ldots,k-1\}, x \in \{0,\ldots,s+n\}$. The calculation is performed using formula (1.12), where the number 1 must be replaced by a number of zero-eigenvalue.

The stationary probability of the ith state of the random environment

$$p_i^J = \sum_{x=0}^{s+n} p_{x,i}, \ i \in \{0,\ldots,k-1\}. \tag{2.10}$$

The stationary intensity of the arrived Poisson flow

$$\alpha St = \sum_{i=0}^{k-1} p_i^J \alpha_i. \tag{2.11}$$

The stationary probability of the rejection

$$P\{Rej\} = \sum_{i=0}^{k-1} \alpha_i \sum_{x=s}^{s+n} p_{x,i}. \tag{2.12}$$

The stationary mean number of working servers

$$E(W) = \sum_{i=0}^{k-1} \left(\sum_{x=1}^{s-1} x p_{x,i} + s \sum_{x=s}^{s+n} p_{x,i} \right). \tag{2.13}$$

The load coefficient of the servers

$$\rho = E(W)/s. \tag{2.14}$$

Example 2.3. The matrix of the transition intensities for Markov chain $J(t)$ is the following:

$$\lambda = \begin{pmatrix} 0 & 0.4 & 1 \\ 0.7 & 0 & 1.1 \\ 0 & 1.5 & 0 \end{pmatrix}.$$

The intensities of the Poisson process $X(t)$ for different states of Markov chain $J(t)$ are given by vector $\alpha = (\alpha_0, \alpha_1, \alpha_2) = (1.5, 2.15, 3)$. The number of s servers equals 4. The service intensities by one server for various states of the process $J(t)$ are defined by the vector $\mu = (\mu_0, \mu_1, \mu_2) = (0.5, 0.75, 1)$.

At the beginning, we consider a case in which the waiting room is absent ($n = 0$); it is a system with rejections.

The number of states $k\{s+1\}$ equals 15. The ordered number of states (i, s) is given in Table 2.2.

These data allow a calculation of the generator G. The corresponding matrix of eigenvectors is presented in Table 2.3. The vector of eigenvalue is as follows:

$$\chi = (-12.16 - 8.91 - 8.69\ 0 - 6.13 - 6.20 - 6.29 - 1.21 - 4.33$$
$$- 4.26 - 4.14 - 1.99 - 2.87 - 2.60 - 2.71).$$

We see that the zero-eigenvalue has the number 3.

Table 2.2. States enumeration $Num(i, x)$.

$x\backslash i$	0	1	2
0	0	1	2
1	3	4	5
2	6	7	8
3	9	10	11
4	12	13	14

Let us present the main criteria of the considered system with rejection for a stationary case. The stationary probability of states

Table 2.3. Matrix B of the eigenvectors.

	0	1	2	3	4
0	−0.016	−0.049	0.098	−0.258	0.22
1	−0.034	0.049	0.206	−0.258	0.382
2	0.103	0.126	−0.477	−0.258	−0.745
3	0.037	0.1	−0.117	−0.258	−0.079
4	0.082	−0.161	−0.243	−0.258	−0.077
5	−0.245	−0.21	0.563	−0.258	0.213
6	−0.071	−0.168	0.069	−0.258	−0.052
7	−0.158	0.374	0.128	−0.258	−0.165
8	0.469	0.278	−0.318	−0.258	0.242
9	0.097	0.211	0.036	−0.258	0.012
10	0.219	−0.578	0.116	−0.258	$-1.484 \cdot 10^{-3}$
11	−0.642	−0.27	−0.207	−0.258	−0.029
12	−0.063	−0.131	−0.065	−0.258	0.058
13	−0.146	0.407	−0.175	−0.258	0.184
14	0.419	0.138	0.342	−0.258	...

	5	6	7	8	9
0	0.325	0.176	0.559	−0.647	−0.651
1	−0.077	$7.612 \cdot 10^{-3}$	0.465	−0.118	0.198
2	−0.59	−0.132	0.371	0.163	0.586
3	−0.301	−0.313	0.258	0.542	0.145
4	0.276	$1.914 \cdot 10^{-3}$	0.221	0.149	−0.116
5	0.373	0.075	0.175	0.068	−0.051
6	0.133	0.494	0.015	−0.208	0.2
7	−0.259	0.059	0.023	−0.145	−0.114
8	−0.029	0.023	0.016	−0.102	−0.159
9	0.086	−0.611	−0.162	−0.194	$-7.632 \cdot 10^{-3}$
10	−0.13	−0.197	−0.122	−0.059	0.034
11	−0.122	−0.078	−0.1	−0.046	−0.028
12	−0.12	0.381	−0.26	0.246	−0.191
13	0.306	0.185	−0.203	0.188	0.166
14	0.042	0.044	−0.164	0.083	...

(*Continued*)

Table 2.3. (*Continued*)

	10	11	12	13	14
0	0.268	0.365	0.688	0.315	0.166
1	0.431	0.08	−0.056	0.13	0.261
2	−0.655	−0.246	−0.45	0.474	−0.323
3	0.1	0.365	0.329	−0.309	0.166
4	0.209	0.08	−0.022	−0.27	0.261
5	−0.293	−0.246	−0.217	0.22	−0.323
6	$1.081 \cdot 10^{-3}$	0.365	0.037	−0.323	0.166
7	0.022	0.08	$4.971 \cdot 10^{-3}$	−0.308	0.261
8	−0.019	−0.246	−0.029	0.182	−0.323
9	−0.06	0.365	−0.176	−0.091	0.166
10	−0.124	0.08	0.024	−0.193	0.261
11	0.17	−0.246	0.107	0.238	−0.323
12	−0.092	0.365	−0.294	0.113	0.166
13	−0.208	0.08	0.034	−0.084	0.261
14	0.271	−0.246	0.182	0.295	...

is as follows:

$$\text{Pst:} = (0.012\ \ 0.025\ \ 0.026\ \ 0.036\ \ 0.074\ \ 0.077\ \ 0.054\ \ 0.108\ \ 0.115\ \ 0.053\ \ 0.106\ \ 0.113\ \ 0.039\ \ 0.077\ \ 0.084)$$

The stationary probabilities of states of the random environment

$$(p_0^J, p_1^J, p_2^J) = (0.194, 0.390, 0.416).$$

The stationary intensity of the arrived Poisson flow (formula (2.11))

$$\alpha St = 2.375.$$

The stationary probability of the rejection (formula (2.12))

$$P\{Rej\} = 0.2.$$

The stationary mean number of working servers (formula (2.13))

$$E(W) = 2.357.$$

The load coefficient of the servers (formula (2.14))

$$\rho = E(W)/s = 0.589.$$

Further, we present the final results for a case in which the waiting room has two places: $n = 2$.

The stationary probability of states is as follows:

$$\text{Pst} := (0.01\ \ 0.02\ \ 0.021\ \ 0.029\ \ 0.059\ \ 0.061\ \ 0.043\ \ 0.086\ \ 0.091\ \ 0.042\ \ 0.084\ \ 0.09\ \ 0.031\ \ 0.062\ \ 0.067\ \ 0.023\ \ 0.045\ \ 0.049\ \ 0.017\ \ 0.033\ \ 0.037)$$

The stationary probabilities of states of the random environment and the stationary intensity of the arrived Poisson flow remain the same.

The stationary probability of the rejection is $P\{Rej\} = 0.087$. The stationary mean number of working servers is $E(W) = 2.693$. The load coefficient of the servers is $\rho = E(W)/s = 0.673$.

We see that the indices of efficiency sufficiently increase.

References

[1] Fischer, W., Meier-Hellstern, K. The Markov modulated Poisson process cookbook. *Performance Evaluation*, 1992;18:149–171.

[2] Kijima, M., *Markov Processes for Stochastic Modeling*, The University Press, Cambridge, UK, 1997.

[3] Özekici, S. Complex systems in random environments. In: S. Özekici (ed.), *Reliability and Maintenance of Complex Systems*, NATO ASI Series. Series F; Computer and Sciences, Vol. 154, Heidelberg, Berlin, Springer-Verlag, New York, 1996, pp. 137–157.

[4] Özekici, S., Soyer, R. Reliability modeling and analysis under random environments. In: Soyer, R., Mazzuchi, T.A., Singpurwalla N.D. (eds.), *Mathematical Reliability: An Expository Perspective.* Kluwer; Boston (MA), 2004, pp. 249–273.

[5] Özekici, S., Soyer, R. Semi-Markov modulated Poisson process: Probabilistic and statistical analysis. *Mathematical Methods of Operations Research*, 2006;64:125–144.

[6] Pacheco, A., Tang, L.C., Prabhu, N.U., *Markov-Modulated Processes & Semiregenerative Phenomena.* World Scientific, New Jersey, London, 2009.

[7] Rydén, T. Parameter estimation for Markov modulated Poisson processes. *Communications in Statistics. Stochastic Models*, 1992; 10:795–829.

[8] Saaty, T.L. *Elements of Queueing Theory with Applications.* McGraw Hill Book Company, INC, New York, Toronto, London, 1961.

[9] Brockmeyer, E. *et al.*, *The Life and Works of A.K. Erlang.* The Copenhagen Telephone Company, Copenhagen, 1948.

[10] Takacs, I. *Introduction to the Theory of Queues*, Oxford University Press, New York, 1962.

Chapter 3

Shortest Paths in Markov-Modulated Networks

3.1 Positing the problem

In this chapter, we consider a *network* $N = (V, E)$ defined by set $V = (v_0, v_2, \dots, v_n)$ of *vertices* and set $E = (e_0, e_1, \dots, e_m)$ of directed *arcs*. Each arc has the form $e = (v_w, v_u)$, where $w < u$. Two vertices are distinguished: *source* v_0 and *target* v_n. It is assumed that every two distinct vertices have no more than one joint arc, and the arc e_η has a constant length $l_\eta > 0$, $\eta = 0, 1, \dots, m$.

The network operates in a random environment described by an irreducible continuous-time Markov chain $J(t), t > 0$ (see Chapter 2). This chain has a finite number k of states. The transition intensities between the states are given by matrix $\lambda = (\lambda_{i,j})$. If the Markov chain has state $i \in \{1, 2, \dots, k\}$, then the speed of transition on arc e_η equals $s_{i,\eta}$.

We exit the source and move from vertex to vertex along the arcs until we reach the target. After entering a vertex, the current state of the random environment is obtained. Using this information, we must select an arc for further transitions.

Our final goal is to reach the target within the minimum possible time. This time is a random value; therefore, we refer to it as the mean time. The corresponding path is referred to as *the shortest path.*

The shortest path problem is one of the classical problems in graph theory [1,2]. We consider a case in which the travel speeds along the arcs are dependent random variables [3–8].

Further expositions are organized as follows: Section 3.2 is devoted to a single arc of the network. The entire network is discussed in Section 3.3. The stationary probability distribution is discussed in Section 3.4. A numerical example is presented in Section 3.1. Sections 3.5–3.2 are dedicated to a particular case: the shortest path in a network containing non-operating arcs. Algorithms and illustrative examples are provided in this study. Section 3.9 contains concluding remarks. Reasoning regarding the computational procedure is presented in the Appendix.

3.2 Travel along a single arc

In this section, we calculate the mean travel time along a fixed arc e_η if the initial state of the random environment is known. It should be noted that the transition intensity of the random environment from state i to state j is $\lambda_{i,j}$. Let Λ_i be the intensity of the transition from state i, $\Lambda_i = \sum_j \lambda_{i,j}$.

Let i be the initial state of the random environment. First, we wish to calculate a probability $P(d)_{i,j}$ that the random environment will have state j at the moment when a length d is covered along the arc with the number η.

The *delta method* is used below. Let $\Delta > 0$ be a sufficiently short length. Furthermore, we neglect the value of the order $o(\Delta)$. Using $n\Delta$ instead of d and $\Pr(\Delta, n)_{i,j}$ instead of $P(d)_{i,j}$, we have for $n = 1$

$$\Pr(\Delta, 1)_{i,j} = \begin{bmatrix} 1 - \dfrac{\Delta}{s_{i,\eta}}\Lambda_i, & j = i, \\ \\ \dfrac{\Delta}{s_{i,\eta}}\lambda_{i,j}, & j \neq i. \end{bmatrix} \tag{3.1}$$

Now, we have a discrete-time finite Markov chain with transition probabilities during one step (3.1). Therefore, we have the following representation of the matrix of transition probabilities $\Pr(\Delta, n) = (\Pr(\Delta, n)_{i,j})$:

$$\Pr(\Delta, n) = \Pr(\Delta, n-1)\Pr(\Delta, 1), \quad n = 2, 3, \ldots. \tag{3.2}$$

This result allows us to calculate the mean number of appearances of state j during n steps if state i occurs initially, denoted $Once(\Delta, n)_{i,j}$. The corresponding matrix is calculated as

$$Once(\Delta, n) = \sum_{\zeta=1}^{n} \Pr(\Delta, \zeta), \quad n = 1, 2, \ldots \tag{3.3}$$

Now, we can calculate the mean time $\mu(\Delta, n)_{i,\eta}$ to cover the distance $n\Delta$ on the arc e_η if the initial state of the random environment is i. Let us remember that the travel speed on the arc e_η equals $s_{i,\eta}$, if the random environment has state i. Let $\tilde{s}_{i,\eta} = 1/s_{i,\eta}$ and vector $\tilde{s}^{\langle\eta\rangle} = (\tilde{s}_{i,\eta}, \tilde{s}_{i,\eta}, \ldots, \tilde{s}_{i,\eta})^T$.

We have the following expression for the vector $\mu^{\langle\eta\rangle}(\Delta, n) = (\mu(\Delta, n)_{\mathbf{1},\boldsymbol{\eta}}, \mu(\Delta, n)_{\mathbf{2},\boldsymbol{\eta}}, \ldots, \mu(\Delta, n)_{\boldsymbol{k},\boldsymbol{\eta}})$:

$$\mu^{\langle\eta\rangle}(\Delta, n) = \Delta\, Once(\Delta, n)\tilde{s}^{\langle\eta\rangle}. \tag{3.4}$$

3.3 Shortest path in the network

We now consider the entire network. Let us introduce some functions which will be necessary henceforth.

Function arc(v, w) as the output parameter has a sequence number η of arc $(v,\ w)$. This function is defined if $(v, w) \in E$. Additionally, if arc$(v, w) = \eta$, then $e_\eta = (v, w)$. The output parameter of the function out(v) is a vector of the numbers of the final vertices for the arc having v as the initial vertex.

The approach discussed below uses a partition of set V at non-overlapping levels $L_0, L_1, \ldots, L_p$. Each level L_ς contains vertices, for which the longest path from the source contains ς arcs. Note that the end levels have only one vertex, which is the source and target: $L_0 = \{v_0\}$, $L_p = \{v_n\}$.

Algorithms for such partitions are well known [1,2]. We assume that this partition is provided.

We consider the vertices sequentially, starting from the last level L_p and ending at level L_0. For each vertex v_ρ, $\rho = 0, \ldots, n$, if the initial state of the random environment is i, the following values must be calculated: (1) minimal mean time $\boldsymbol{\tau}_{\boldsymbol{i},\boldsymbol{\rho}}$ until the target; and (2) the following vertex $u_{i,\rho} \in$ out(ρ), somewhere it should go from vertex v_ρ.

For the last vertex, $\tau_{i,n} = 0$, $i = 1, \ldots, k$. If vertex $v_\rho \in L_{p-1}$ then $u_{i,\rho} = v_n$, $i = 1, \ldots, k$. The remaining values of $\{\boldsymbol{\tau}_{i,\rho}\}$ and $\{u_{i,\rho}\}$ are calculated with the *following algorithm.*

Step 0. Set $\psi = p - 1$.
Step ψ. Organize a cycle over the vertices v_ρ from L_ψ.

1. For the current vertex v_ρ organize a cycle over outgoing arcs.

For a current arc (v_ρ, v_{ρ^*}) by the number $\eta = \text{arc}(v_\rho, v_{\rho^*})$,

1.1. Define its length $l_\eta = n\Delta$ and speeds $s_{i,\eta}$, $i = 1, \ldots, k$.
1.2. Calculate a vector of the mean times $\mu^{\langle\eta\rangle}(\Delta, n)$ to cover distance $n\Delta$ by formula (3.4).
1.3. Calculate a matrix $\Pr(\Delta, n, \eta)$ of transition probabilities for the states of the random environment.
1.4. Calculate a vector of mean full time $\mu f^{\langle\eta\rangle}$ until the target by formula

$$\mu f^{\langle\eta\rangle} = \mu^{\langle\eta\rangle}(\Delta, n) + \Pr(\Delta, n, \eta)\tau_{\rho^*}, \tag{3.5}$$

where $\tau_{\rho^*} = (\tau_{1,\rho^*}, \ldots, \tau_{k,\rho^*})^T$.
2. Organize a cycle over states of the random environment.
2.1. For the current state i, to define the number η^* of an arc with minimal value of the component $\mu f_i^{\langle\eta\rangle}$.
2.2. To assign $\tau_{i,\rho} = \mu f_i^{\langle\eta^*\rangle}$. Let

$$\tau_\rho = (\tau_{1,\rho}, \ldots, \tau_{k,\rho})^T. \tag{3.6}$$

2.3. If v_{ρ^*} is a final vertex for arc e_{η^*}, that is, $\eta^* = \text{arc}(v_\rho, v_{\rho^*})$, then assign $u_{i,\rho} = v_{\rho^*}$.

3.4 Stationary probability distribution

Now, we consider a stationary probability distribution of random environment states [9,10]. Let $\text{diag}(\Lambda)$ be a diagonal matrix with the vector Λ on the main diagonal. Matrix $A = \lambda - \text{diag}(\Lambda)$ is called a *generator.* Let $\chi_1, \ldots, \chi_k$ be *eigenvalues* and $\beta_1, \ldots, \beta_k$ be corresponding *eigenvectors* of the generator. We assume that all the eigenvalues are different.

Further, let $\chi = (\chi_1, \ldots, \chi_k)$, $B = (\beta_1, \ldots, \beta_k)$, and let the inverse matrix with rows $\tilde{\beta}_1, \ldots, \tilde{\beta}_k$ be $B^{-1} = (\tilde{\beta}_1^T, \ldots, \tilde{\beta}_k^T)^T$. Let $P_{i,j}(t) = P\{J(t) = j[J(0) = i\}$ be the transition probability from state i to state j during time t. The corresponding matrix $P(t) = (P_{i,j}(t))$ is computed using the formula (1.9):

$$P(t) = \sum_{i=1}^{k} \exp(\chi_i t)\beta_i\tilde{\beta}_i = B\,\mathrm{diag}(\exp(\chi t))B^{-1}, \quad t \geq 0, \qquad (3.7)$$

where $\exp(\chi t) = (\exp(\chi_i t) \quad \cdots \quad \exp(\chi_k t))^{\mathrm{T}}$.

We denote the stationary distribution of process $J(t)$ by $\pi = (\pi_1, \ldots, \pi_k)$, where

$$\pi_j = \lim_{t\to\infty} P_{i,j}(t))$$

and it is assumed that this limit exists and does not depend on the initial state i.

The aforementioned is possible if only one of the eigenvalues equals zero, and the others are positive.

Let the zero eigenvalue be enumerated by one. Then

$$\lim_{t\to\infty} P(t) = \lim_{t\to\infty} \sum_{i=1}^{k} \exp(\chi_i t)\beta_i\tilde{\beta}_i$$

$$= \exp(0)\beta_1\tilde{\beta}_1 + \lim_{t\to\infty} \sum_{i=2}^{k} \exp(\chi_i t)\beta_i\tilde{\beta}_i = \beta_1\tilde{\beta}_1. \quad (3.8)$$

It is known that all components of the eigenvector β_1 are equal. Therefore, all the rows of the limit matrix $\lim_{t\to\infty} P(t)$ coincide. This repeated row yields the stationary distribution $\pi = (\pi_1, \ldots, \pi_k)$.

Above, we obtained an expression for the mean time $\mu(\Delta, n)_{i,\eta}$ for overcoming the whole length of the arc e_η, if the initial state of the random environment is i. The unconditional value is the following:

$$\tilde{\mu}(\Delta, n)_\eta = \sum_{i=1}^{k} \pi_i \mu(\Delta, n)_{i,\eta}, \quad \eta \in E. \qquad (3.9)$$

Analogously, we have expression (3.6) for the vector τ_ρ of the mean full-time to the target from vertex ρ. This vector contains the values for the different initial states of a random environment. Averaging over the state distribution gives the unconditional mean full-time to the target from different vertices:

$$\tilde{\tau} = \pi \tau_0. \tag{3.10}$$

In conclusion, we provide a short explanation of the importance of having information about the states of the random environment. Let us suppose that such information is missing and that the unconditional mean time (3.9) is available only for single arcs. This case corresponds to the classical problem of shortest paths. The numerical results for these two cases are presented as follows.

Example 3.1. The considered network is defined in Table 3.1. Each column of the table corresponds to an arc. The first element of each column provides the number of arcs, the second element is the number of initial vertices for the arc, and the third element is the number of final vertices.

Lengths of the arcs are given in the second row of Table 3.2.

A random environment has three states. The transition intensities between states $\{\lambda_{i,j}\}$ and the output intensities for states $\{\Lambda_i\}$ are

Table 3.1. Arcs and vertices.

0	1	2	3	4	5	6	7	8	9	10	11
0	0	0	1	1	2	2	3	3	4	5	6
1	2	3	3	6	4	5	4	6	7	7	7

Table 3.2. Lengths of the arcs.

0	1	2	3	4	5	6	7	8	9	10	11
4	5	4	7	11	10	11	9	7	5	6	7

Table 3.3. Speeds $s_{i,\eta}$ of the travel.

0	1	2	3	4	5	6	7	8	9	10	11
1	2	3	1	1	3	2	3	2	3	2	3
2	3	4	2	3	4	5	4	4	5	4	7
5	4	5	6	3	4	6	5	6	5	5	8

as follows:

$$\lambda = \{\lambda_{i,j}\} = \begin{pmatrix} 0 & 0.04 & 0.10 \\ 0.12 & 0 & 0.16 \\ 0 & 0.2 & 0 \end{pmatrix}, \quad \Lambda = \{\Lambda_i\} = \begin{pmatrix} 0.14 \\ 0.28 \\ 0.20 \end{pmatrix}.$$

Table 3.3 contains the speeds $s_{i,\eta}$ of travel on different arcs e_η and different states of random environment i.

Further, computation results are presented.

First, we present, as an example, matrices $\Pr(\Delta, n, \eta)$ for the sixth arc ($\eta = 6$), $d = n\Delta = 11$, and two values of Δ: $\Delta = 0.01$ and $\Delta = 0.05$.

$$\Pr(0.01, 700, 6) = \begin{pmatrix} 0.482 & 0.172 & 0.346 \\ 0.137 & 0.599 & 0.264 \\ 0.028 & 0.232 & 0.740 \end{pmatrix},$$

$$\Pr(0.05, 350, 6) = \begin{pmatrix} 0.481 & 0.172 & 0.347 \\ 0.137 & 0.598 & 0.265 \\ 0.028 & 0.233 & 0.739 \end{pmatrix}.$$

A comparison shows that the value $\Delta = 0.05$ can be used. Additional information about choosing Δ can be found in the Appendix.

Table 3.4 contains the mean times $\mu(\cdots)_{i,\eta}$ to cover the full length of the arc e_η if the initial state of the random environment is i. The first row contains the number of arcs and the following rows correspond to the states of the random environment.

The next step is to partition the considered network into the following levels:

$$L_0 = \{0\}, \quad L_1 = \{1, 2\}, \quad L2 = \{3\}, \quad L_4 = \{4, 5, 6\}, \quad L_5 = \{7\}.$$

The final results are presented in Tables 3.5 and 3.6.

Table 3.4. Mean time $\mu(\ldots)_{i,\eta}$ for the ηth arc and initial state i.

$\eta=1$	$\eta=2$	$\eta=3$	$\eta=4$	$\eta=5$	$\eta=6$	$\eta=7$	$\eta=8$	$\eta=9$	$\eta=10$	$\eta=11$	$\eta=12$
3.367	2.348	1.307	5.178	7.548	3.185	4.462	2.824	3.072	1.610	2.709	2.139
2.026	1.701	1.015	3.401	4.454	2.601	2.442	2.272	1.828	1.045	1.586	1.065
0.905	1.314	0.826	1.426	3.870	2.527	1.931	1.885	1.245	1.012	1.250	0.895

Table 3.5. Matrix of the shortest times $(\tau_{i,\rho})$.

i/ρ	0	1	2	3	4	5	6
0	5.245	5.480	4.007	4.240	1.610	2.709	2.139
1	4.021	2.578	3.199	2.996	1.045	1.586	1.065
2	2.998	1.981	3.060	2.189	1.012	1.250	0.895

Table 3.6. Matrix of the optimal following vertices $(u_{i,\rho})$.

i/ρ	0	1	2	3	4	5	6
0	2	6	4	4	7	7	7
1	2	6	4	6	7	7	7
2	2	6	4	6	7	7	7

Next, we consider the stationary case. Generator $A = \lambda - \text{diag}(\Lambda)$ is expressed as follows:

$$A = \begin{pmatrix} -0.14 & 0.04 & 0.10 \\ 0.12 & -0.28 & 0.16 \\ 0 & 0.2 & -0.2 \end{pmatrix}.$$

The vector of eigenvalues $\chi = (\chi_1, \ldots, \chi_k)$, matrix of eigenvectors $B = (\beta_1, \ldots, \beta_k)$, and inverse matrix B^{-1} are presented as follows:

$$\chi = \begin{pmatrix} -0.408 \\ -0.212 \\ 0 \end{pmatrix}, \quad B = \begin{pmatrix} -0.149 & -0.806 & 0.577 \\ -0.714 & -0.034 & 0.577 \\ 0.085 & 0.590 & 0.577 \end{pmatrix},$$

$$B^{-1} = \begin{pmatrix} 0.436 & -0.975 & 0.539 \\ -0.976 & 0.582 & 0394 \\ 0.481 & 0.561 & 0.690 \end{pmatrix}.$$

Table 3.7. Mean time $\tilde{\mu}(\Delta, n)_\eta$ for the stationary case.

$\eta = 0$	$\eta = 1$	$\eta = 2$	$\eta = 3$	$\eta = 4$	$\eta = 5$	$\eta = 6$	$\eta = 7$	$\eta = 8$	$\eta = 9$	$\eta = 10$	$\eta = 11$
1.952	1.726	1.021	3.109	5.081	2.734	2.800	2.271	1.941	1.189	1.764	1.296

Table 3.8. Unconditional mean full-time $\tilde{\tau}_\rho$ to the target from different vertices.

$\rho = 0$	$\rho = 1$	$\rho = 2$	$\rho = 3$	$\rho = 4$	$\rho = 5$	$\rho = 6$
4.009	5.631	3.3799	3.021	1.189	1.764	1.296

This allows us to calculate the transition probabilities (3.7) and stationary probabilities (3.8) for the states of the random environment. In particular, the stationary probabilities

$$\pi = (\pi_1, \ldots, \pi_k) = \begin{pmatrix} 0.278 & 0.324 & 0.398 \end{pmatrix}.$$

Table 3.7 shows the unconditional mean time $\tilde{\mu}(\Delta, n)_\eta$ calculated using formula (3.9) for the stationary case.

Now, formula (3.10) gives the unconditional mean full-time to the target from different vertices. The times are listed in Table 3.8.

In addition, the minimal unconditional mean time from source to target equals 4.009. It is interesting to compare this with the case in which information about the state of the random environment is not used. In this case, we are guided by the mean time $\tilde{\mu}(\Delta, n)_\eta$ to cover arc e_η. The corresponding values are presented in Table 3.7. It is easy to see that the shortest path passes through vertices v_0, v_3, v_4, v_7 or arcs e_2, e_7, e_9. Table 3.7 shows that this path has the following length:

$$1.021 + 2.271 + 1.189 = 4.481.$$

We see the result is worse at 100% $(4.481 - 4.009)/4.009 = 11.8\%$.

3.5 Network containing non-operating arcs

In this section, we introduce the following changes to the problem described in Section 3.1. We suppose that an arc e_η does not operate if the ith state of the random environment occurs. This means that the distance covered on arc e_η does not change while the state i takes place, so $s_{i,\eta} = 0$. Let $L(i) = \{\eta \in V : s_{i,\eta} = 0\}$ be the set of such arcs for state i. We assume that there exists a state i for each arc η with $s_{i,\eta} > 0$.

The task is to determine *the shortest path* in the network for these conditions. We use the known results of the transition probabilities of the states of continuous-time finite ergodic Markov chains [9–11]. Therefore, as stated previously in Section 3.4, $\Lambda_i = \sum_{\varsigma=1}^{k} \lambda_{i,\varsigma}$, $\Lambda = (\Lambda_1, \ldots, \Lambda_k)^T$ and $\text{diag}(\Lambda)$ is a diagonal matrix with vector Λ on the main diagonal. Matrix $A = \lambda - \text{diag}(\Lambda)$ is called a *generator.* Let $\chi_1, \ldots, \chi_k$ be the *eigenvalues* and $\beta_1, \ldots, \beta_k$ be corresponding *eigenvectors* of the generator. We assume that all the eigenvalues are different.

Furthermore, let $\chi = (\chi_1, \ldots, \chi_k)$, $B = (\beta_1, \ldots, \beta_k)$, and the inverse matrix with rows $\bar{\beta}_1, \ldots, \bar{\beta}_k$ be $B^{-1} = (\bar{\beta}_1^T, \ldots, \bar{\beta}_k^T)^T$, and $\text{diag}(\exp(t\chi))$ is the diagonal matrix with the vector $\exp(t\chi) = (\exp(t\chi_1), \ldots, \exp(t\chi_k))$ on the main diagonal. Let $P_{i,j}(t) = P\{J(t) = j \,|\, J(0) = i\}$ be a transition probability from state i to state j during time t. The corresponding matrix $P(t) = (P_{i,j}(t))$ is calculated as follows:

$$P(t) = \sum_{i=1}^{k} \exp(\chi_i t)\beta_i \tilde{\beta}_i = B\,\text{diag}(\exp(\chi t))B^{-1}, \quad t \geq 0, \qquad (3.11)$$

where $\exp(\chi t) = (\exp(\chi_i t) \quad \cdots \quad \exp(\chi_k t))^{\mathrm{T}}$.

This expression is fundamental to the calculations. Usually, the matrix exponent is used instead of (3.11) [12]. The last is presented by the infinity matrix sum, which is not as strong as (3.11) in terms of analytical and computational aspects.

In this section, we aim to calculate the mean travel time along a fixed arc $\boldsymbol{e}_\eta$, assuming that the initial state of the random environment is known.

First, we present the simplest approach to this calculation. The mean time when Markov chain $J(t)$ is in state j on the interval $(0, t)$ equals $\int_0^t \boldsymbol{P}_{i,j}(\boldsymbol{z})d\boldsymbol{z}$. The corresponding mean covered distance $X(t)$ equals $\boldsymbol{s}_{j,\eta} \int_0^t \boldsymbol{P}_{i,j}(\boldsymbol{z})d\boldsymbol{z}$. Therefore, the mean covered distance on the interval $(0, t)$ is calculated as follows:

$$E(X(t) \mid J(0) = i) = \sum_{j \in I(\eta)} s_{j,\eta} \int_0^t P_{i,j}(z) dz, \tag{3.12}$$

where $I(\eta) = \{j \in \{1, \ldots, k\} : s_{j,\eta} > 0\}$ is the set of states when arc e_η operates; we refer to these states as *operational* for this arc.

Let $t^*(i, \eta)$ be the mean travel time along arc e_η if the initial state is i. Because the length of this arc equals d_η, then

$$t^*(i, \eta) = \inf(t : E(X(t) \mid J(0) = i) \geq d_\eta) \tag{3.13}$$

The last two formulas do not consider the following: each distance is reached at the instant when a state with positive speed takes place. The mean covered distance on the interval $(0, t)$, together with the probability that a final state belongs to set $I(\eta)$, is as follows:

$$\sum_{j \in I(\eta)} s_{j,\eta} \int_0^t P_{i,j}(z) \sum_{v \in I(\eta)} P_{j,v}(t - z) dz.$$

The unconditional expectation is calculated as usual:

$$\widetilde{E}(X(t) \mid J(0) = i) = \left(\sum_{v \in I(\eta)} P_{i,v}(t) \right)^{-1} \sum_{j \in I(\eta)} s_{j,\eta} \int_0^t P_{i,j}(z) \times \sum_{v \in I(\eta)} P_{j,v}(t - z) dz.$$

Now instead of (3.13), we have

$$t^*(i, \eta) = \inf(t : \widetilde{E}(X(t) | J(0) = i) \geq d_\eta). \tag{3.14}$$

The procedure presented above can be improved for arcs with all operational states when $I(\eta) = \{1, \ldots, k\}$. Let us substitute time t as an argument of process $J(t)$ by the covered distance x to obtain the process $\widetilde{J}(x)$. Let $X(t)$ be the distance covered at time t along the considered arc:

$$X(t) = \int_0^t s_{J(z),\eta} dz, \quad t \geq 0.$$

Then,

$$\widetilde{J}(x) = J(\inf(t : X(t) \geq x) \,|\, X(0) = i), \quad x \geq 0.$$

Let $\widetilde{P}_{i,j}(x)$ be the probability of state j of random environment $J(t)$ at time t when $X(t) = x$:

$$\widetilde{P}_{i,j}(x) = P\{\widetilde{J}(x) = j \,|\, \widetilde{J}(0) = i\} = P_{i,j}(\inf(t : X(t) \geq x)), \quad x \geq 0.$$

If the random environment has the ith state, then the increment Δx of the covered distance during time Δt equals $\Delta t s_{i,\eta}$. Therefore, the distance Δx is covered during time $\Delta t = \Delta x / s_{i,\eta}$. The usual reasoning gives the following result for a small increment $\Delta x > 0$:

$$\widetilde{P}_{i,j}(x + \Delta x) = \widetilde{P}_{i,j}(x) \left(1 - \frac{\Delta x}{s_{i,\eta}} \sum_{\nu \neq i} \lambda_{i,\nu}\right)$$

$$+ \sum_{\nu \neq i} \widetilde{P}_{\nu,j}(x) \frac{\Delta x}{s_{i,\eta}} \lambda_{i,\nu} + o(\Delta x), \quad x \geq 0,$$

$$\frac{\partial}{\partial x} \widetilde{P}_{i,j}(x) = -\widetilde{P}_{i,j}(x) \frac{1}{s_{i,\eta}} \Lambda_i + \sum_{\nu \neq i} \widetilde{P}_{\nu,j}(x) \frac{1}{s_{i,\eta}} \lambda_{i,\nu}, \quad x \geq 0.$$

Using matrix $\widetilde{P}(x) = (\widetilde{P}_{i,j}(x))$ and vectors $s_\eta = (s_{1,\eta}, \ldots, s_{k,\eta})$, $\Lambda = (\Lambda_1, \ldots, \Lambda_k)$, we can write

$$\frac{\partial}{\partial x} \widetilde{P}(x) = -\text{diag}(s)^{-1} \text{diag}(\Lambda) \widetilde{P}(x) + \text{diag}(s_\eta)^{-1} \lambda \widetilde{P}(x)$$

$$= \text{diag}(s_\eta)^{-1} (\lambda - \text{diag}(\Lambda)) \widetilde{P}(x), \quad x \geq 0.$$

We observe that the substitution of time t by the covered distance x leads to a change of the generator $\lambda - \text{diag}(\Lambda)$ by the generator $\text{diag}(s_\eta)^{-1}(\lambda - \text{diag}(\Lambda))$. This allows us to use formula (3.11) for the calculation of the matrix $\widetilde{P}(x)$ after the corresponding changes of

$\chi = (\chi_1, \ldots, \chi_k)$, $B = (\beta_1, \ldots, \beta_k)$, and inverse matrix B^{-1} with rows $\bar{\beta}_1, \ldots, \bar{\beta}_k$.

Now, the mean travel time along the ηth link if the ith state was initial is calculated in such a way as illustrated by formula (3.13):

$$t^*(i, \eta) = \sum_{j \in I(\eta)} (s_{j,\eta})^{-1} \int_0^{d_\eta} \widetilde{P}_{i,j}(x) dx. \tag{3.15}$$

3.6 Probabilities of final states during travel along a single arc

The formulas presented above allow the calculation of the mean time to cover for any initial state and arc. The algorithm used for the shortest path must consider the probabilities of the final states for different arcs. If a considered arc is operational for all states, then these probabilities are presented by matrix $\widetilde{P}(x)$ as described above, with $x \doteq d_\eta$ for the ηth arc.

Now, we consider an arc e_η, which has nonoperational states. In this case, it is only of interest to consider the states $I(\eta)$ when the arc is operational. Let us calculate the transient intensities between the states of set $I(\eta)$ during a time when the considered arc e_η is operational. Let $\tilde{\lambda} = (\tilde{\lambda}_{i,j})$ be corresponding matrix for states $\{i, j \in I(\eta)\}$.

Let $\bar{I}(\eta) = \{1, \ldots, k\} - I(\eta)$ and $\rho_{i,j}$ be the probability that process $J(t)$ will be in state $j \in I(\eta)$ after leaving the set $\bar{I}(\eta)$, if the initial state is $i \in \bar{I}(\eta)$. Obviously,

$$\rho_{i,j} = \frac{\lambda_{i,j}}{\Lambda_i} + \sum_{\nu \in \bar{I}(\eta)} \frac{\lambda_{i,\nu}}{\Lambda_i} \rho_{\nu,j}. \tag{3.16}$$

Further, we use the matrix $R = (\rho_{i,j})$, the sub-matrices $\bar{\lambda} = (\lambda_{i,j})$, $\hat{\lambda} = (\lambda_{i,\nu})$ of the matrix λ, and the vector $\bar{\Lambda} = (\Lambda_i)$ with $\nu \in \bar{I}(\eta)$, $j \in I(\eta)$. Explicitly,

$$R = \text{diag}(\bar{\Lambda})^{-1}\bar{\lambda} + \text{diag}(\bar{\Lambda})^{-1}\hat{\lambda}R.$$

Therefore,

$$R = (E - \text{diag}(\bar{\Lambda})^{-1}\hat{\lambda})^{-1}\text{diag}(\bar{\Lambda})^{-1}\bar{\lambda}, \tag{3.17}$$

where E is a unit matrix of the corresponding dimension.

Now we can calculate intensities $\tilde{\lambda}_{i,j}$ for $i, j \in I(\eta)$ and matrix $\tilde{\lambda} = (\tilde{\lambda}_{i,j})$:

$$\tilde{\lambda}_{i,j} = \lambda_{i,j} + \sum_{\nu \in \overline{I}(\eta)} \lambda_{i,\nu} R_{\nu,j}, \quad i, j \in I(\eta).$$

Let us consider two sub-matrices of the initial transition intensities $\{\lambda_{i,j}\}$ between states: (1) the sub-matrix $\lambda^* = (\lambda_{i,j})$ of the matrix for transitions between states of the set $I(\eta)$; and (2) the sub-matrix $\lambda^{**} = (\lambda_{i,\nu})$ for transitions from states of set $I(\eta)$ to states of set $\overline{I}(\eta)$. Then,

$$\tilde{\lambda} = \lambda^* + \lambda^{**} R. \tag{3.18}$$

Therefore, we need to exchange matrix λ using matrix $\text{diag}\left(\overline{\overline{s}}^{\langle \eta \rangle}\right)^{-1} \tilde{\lambda}$, where $\overline{\overline{s}}^{\langle \eta \rangle}$ is the sub-vector of the vector $\text{s}^{\langle \eta \rangle} = (\text{s}_{1,\eta}, \text{s}_{2,\eta}, \ldots, \text{s}_{k,\eta})^T$ with positive components. This allows us to use formula (3.11) to calculate matrix $\widetilde{P}(x)$ after the corresponding changes of $\chi = (\chi_1, \ldots, \chi_k)$, $B = (\beta_1, \ldots, \beta_k)$, and inverse matrix B^{-1} with rows $\bar{\beta}_1, \ldots, \bar{\beta}_k$.

In addition, the matrix $\widetilde{P}(d_\eta)$ provides the probabilities of the final states for the set $I(\eta)$. If $j \in \overline{I}(\eta)$, then the corresponding final probability is zero.

Now, we consider the last possible case: the initial state i belongs to set $\overline{I}(\eta)$, the final state j belongs to set $I(\eta)$. We denote the corresponding as $\widetilde{\Pr}(\eta)$. Let us recall that $\rho_{i,j}$ is the probability that chain $J(t)$ will be in state $j \in I(\eta)$ after leaving the set $\overline{I}(\eta)$, if the initial state is $i \in \overline{I}(\eta)$. Therefore,

$$\widetilde{\Pr}(\eta) = R\widetilde{P}(d_\eta). \tag{3.19}$$

All possible cases are considered. Let $\Pr(\eta)$ be the total notation of the considered probabilities of the final state of arc e_η.

3.7 Shortest path for the network with non-operating arcs

We now consider the entire network, which includes many arcs. We use the functions $\text{arc}(v, w)$ and $\text{out}(v)$, as defined in Section 3.3. We also assume that the partition of the vertex set V in non-overlapping levels $L_0, L_1, \ldots, L_p$ is given. The vertices are considered sequentially, beginning from the last level L_p and ending at level L_0. We set $L_0 = \{v_0\}$ and $L_p = \{v_n\}$, respectively. An intermediate level L_w has the following property: each of its vertices has several paths until the target v_n, containing nodes from levels with larger numbers only. Our aim is to calculate the following:

1. Minimal mean time $\boldsymbol{\tau_{i,\rho}}$ until the target for vertex $v_\rho \in V$, if the initial state of the random environment is i. Let

$$\tau_\rho = (\tau_{1,\rho}, \ldots, \tau_{k,\rho})^T$$

be the corresponding vector-column.

2. The following vertex $\boldsymbol{u_{i,\rho}} \in \boldsymbol{out(\rho)}$ should be reached from vertex v_ρ, $\rho = 0, \ldots, n-1$. If vertex $\boldsymbol{v_\rho} \in \boldsymbol{L_{p-1}}$, then $\boldsymbol{u_{i,\rho}} = \boldsymbol{v_n}$, $\boldsymbol{e_\eta} = (\boldsymbol{v_\rho}, \boldsymbol{v_n})$, $\boldsymbol{\tau_{i,\rho}} = \boldsymbol{t^*(i, \eta)}$. The remaining values of $\{\boldsymbol{\tau_{i,\rho}}\}$ and $\{\boldsymbol{u_{i,\rho}}\}$ are calculated with respect to *the following algorithm.*

Step 0. Organize a loop over the levels L_ψ, $\psi = p-2, \ldots, 0$.
Step ψ. Organize a cycle over the vertices v_ρ from L_ψ.

Use the current vertex v_ρ to organize a cycle over the output arcs $\eta = (v_\rho, v_{\rho^*})$.

1. Calculate for a current arc from vertex ρ to vertex ρ^* by number $\eta = \text{arc}(v_\rho, v_{\rho^*})$ and all initial states i,

1.1. Calculate the vector $\boldsymbol{t^*(\eta) = (t^*(1, \eta), \ldots, t^*(k, \eta))^T}$ of the mean times, which are necessary to cover the distance and length of the considered arc (see formulas (3.14) and (3.15)).

1.2. Calculate probability $\mathbf{Pr}_{\boldsymbol{i,j}}(\boldsymbol{d_\eta}), \boldsymbol{i, j} = \mathbf{1}, \ldots, \boldsymbol{k}$, that the random environment will have state j at the instant when the distance/length $\boldsymbol{d_\eta}$ is covered. Let $\mathbf{Pr}(\boldsymbol{\eta}) = (\mathbf{Pr}_{\boldsymbol{i,j}}(\boldsymbol{d_\eta}))$ be the corresponding $(k \times k)$-matrix.

1.3. Calculate the vector of the mean full times $\hat{\boldsymbol{t}}_{\boldsymbol{i,\eta}}$ up to the target, if the arc η is used:

$$\hat{t}_\eta = (\hat{t}_{0,\eta}, \ldots, \hat{t}_{k-1,\eta})^T = t^*(\eta) + \Pr(\eta)\text{diag}(\tau_{\rho^*}). \qquad (3.20)$$

2. Organize a cycle over the arcs from the vertex $\boldsymbol{v_\rho}$.

2.1. For current state i, define number η^* of arcs with minimal value of $\hat{\boldsymbol{t}}_{\boldsymbol{i,\eta}}$.

2.2. Assign $\boldsymbol{\tau_{i,\rho}} = \hat{\boldsymbol{t}}_{i,\eta^*}$. Let

$$\tau_\rho = (\tau_{1,\rho}, \ldots, \tau_{k,\rho})^T. \quad (3.21)$$

2.3. If v_{ρ^*} is an end vertex of link e_{η^*}, that is, $\eta^* = \mathrm{arc}(v_\rho, v_{\rho^*})$, then assign $\boldsymbol{u_{i,\rho}} = \boldsymbol{v_{\rho^*}}$.

3.8 Stationary probability distribution for the network with non-operating arcs

We now consider the stationary probability distribution of random environment states. We denote the stationary distribution of process $J(t)$ by $\pi = (\pi_1, \ldots, \pi_k)$, where

$$\boldsymbol{\pi}_j = \lim_{t\to\infty} \boldsymbol{P}_{i,j}(\boldsymbol{t}).$$

It is assumed that only one eigenvalue of generator A equals zero, and the other eigenvalues are negative. Let the zero eigenvalue be number i^*, $\boldsymbol{\beta}_{i^*}$ be its eigenvector, and $\tilde{\beta}_{i^*}$ be the i^*th row of the inverse matrix B^{-1}. Then,

$$\lim_{t\to\infty} P(t) = \beta_{i^*}\bar{\beta}_{i^*}. \quad (3.22)$$

All the components of the eigenvector $\boldsymbol{\beta}_{l^*}$ are equal. Therefore, all rows of the limit matrix $\mathbf{lim}_{t\to\infty}\boldsymbol{P}(\boldsymbol{t})$ coincide. This repeated row yields a stationary distribution $\pi = (\pi_1, \ldots, \pi_k)$.

Now, we must average the indices of interest. For example, the stationary mean time $\boldsymbol{\mu s}(\boldsymbol{\eta})$ of the passing arc e_η can be calculated as follows:

$$\mu s(\eta) = (\pi_1, \ldots, \pi_k)(t^*(1,\eta), \ldots, t^*(k,\eta))^T, \quad \eta = 0, \ldots, m. \quad (3.23)$$

The stationary minimal mean time $\boldsymbol{\tau s_\rho}$ until the target for vertex $v_\rho \in V$ is

$$\boldsymbol{\tau s_\rho} = \boldsymbol{\pi}\,\boldsymbol{\tau_\rho} = (\boldsymbol{\pi_1}, \ldots, \boldsymbol{\pi_k})(\boldsymbol{\tau_{1,\rho}}, \ldots, \boldsymbol{\tau_{k,\rho}})^{\boldsymbol{T}}.$$

In particular, the stationary mean time for the shortest path from source v_0 to target v_n equals $\boldsymbol{\tau s_0}$.

Example 3.2. Let us continue the analysis of the network defined in Tables 3.1 and 3.2 in Section 3.4. Table 3.9 contains the speeds $s_{i,\eta}$ of travel along the arcs e_η for different states of random environment i.

The random environment has three states as before (see Section 3.4). Transition intensities between the states $\{\lambda_{i,j}\}$, output intensities for the states $\{\Lambda_i\}$, as well as the main characteristics of the Markov chain: generator A, vector of eigenvalues χ, matrix of eigenvectors B, and inverse matrix B^{-1} are the same as in Section 3.4.

Now, it is possible to calculate the transition probabilities $\boldsymbol{P_{i,j}(t) = P\{J(t)\,|\,J(0) = i\}}$ of the Markov chain as a function of time t (see formula (2.11)). This allows the calculation of the mean time $\boldsymbol{t^*(i,\eta)}$ of the travel for arcs e_4 and e_5 with non-operational states if the initial state is i (see formula (3.14)). The mean time $\boldsymbol{t^*(i,\eta)}$ for the rest of the arcs is calculated using formula (3.15). The results are presented in Table 3.10.

Furthermore, the probabilities of the final states for different arcs are presented. First, we consider the 6th arc, which is operational for

Table 3.9. Speeds $s_{i,\eta}$ of travel.

0	1	2	3	4	5	6	7	8	9	10	11
10	5	3.33	10	0	3.33	5	3.33	5	3.33	5	3.22
5	3.33	2.5	5	3.33	0	2	2.5	2.5	2	2.5	1.43
2	2.5	2	1.67	0	2.5	1.67	2	1.67	2	2	1.25

Table 3.10. Mean time $t^*(i,\eta)$ for the ηth arc and initial state i.

$\eta=0$	$\eta=1$	$\eta=2$	$\eta=3$	$\eta=4$	$\eta=5$	$\eta=6$	$\eta=7$	$\eta=8$	$\eta=9$	$\eta=10$	$\eta=11$
0.432	1.055	1.252	0.809	13.40	3.15	2.711	2.933	1.600	1.595	1.322	2.504
0.846	1.510	1.609	1.567	3.28	6.51	5.061	3.627	2.822	2.386	2.346	4.519
1.795	1.913	1.931	3.336	8.80	4.26	5.959	4.201	3.754	2.482	2.842	5.203

all the states. The generator and other attributes for the 6th arc are as follows:

$$\boldsymbol{AD}(6) = \begin{pmatrix} -0.028 & 0.008 & 0.020 \\ 0.060 & -0.140 & 0.080 \\ 0.000 & 0.120 & -0.12 \end{pmatrix}, \quad \chi\mathrm{D}(6) = \begin{pmatrix} -0.226 \\ 0 \\ -0.062 \end{pmatrix},$$

$$\boldsymbol{BD}(6) = \begin{pmatrix} 0.049 & -0.577 & -0.535 \\ 0.662 & -0.577 & 0.368 \\ -0.748 & -0.577 & 0.760 \end{pmatrix},$$

$$\boldsymbol{BD}(6)^{-1} = \begin{pmatrix} -0.259 & 0.856 & -0.597 \\ -0.891 & -0.416 & -0.426 \\ -0.931 & 0.527 & 0.405 \end{pmatrix}.$$

It allows us to calculate the final probabilities of the states $\mathbf{Pr}_{i,j}(\mathbf{6})$. The corresponding matrix is expressed as follows:

$$\mathbf{Pr}(6) = \begin{pmatrix} 0.765 & 0.101 & 0.134 \\ 0.327 & 0.385 & 0.288 \\ 0.172 & 0.389 & 0.439 \end{pmatrix}.$$

As shown in Table 3.9, the two arcs have nonoperational states. Namely, they are the first and third states for the fourth arc and the second state for the fifth arc. Therefore, the sets of states when these arcs operate are as follows: $I(4) = \{2\}$ and $I(5) = \{1, 3\}$. Let us give the transient intensities and the transient probabilities that occur in formulas (3.16)–(3.18).

The sub-matrices $\bar{\boldsymbol{\lambda}} = (\boldsymbol{\lambda}_{i,j})$, $\hat{\boldsymbol{\lambda}} = (\boldsymbol{\lambda}_{i,v})$ of the matrix $\boldsymbol{\lambda}$, and the vector $\bar{\boldsymbol{\Lambda}} = (\boldsymbol{\Lambda}_i)$ with $\boldsymbol{i}, \boldsymbol{\nu} \in \bar{\boldsymbol{I}}(\eta)$, $\boldsymbol{j} \in \boldsymbol{I}(\eta)$, are the following for the fourth and the fifth links:

$$\bar{\boldsymbol{\lambda}}(4) = \begin{pmatrix} 0.04 \\ 0.20 \end{pmatrix}, \quad \hat{\boldsymbol{\lambda}}(4) = \begin{pmatrix} 0 & 0.1 \\ 0 & 0 \end{pmatrix}, \quad \bar{\boldsymbol{\Lambda}}(4) = \begin{pmatrix} 0.14 \\ 0.20 \end{pmatrix},$$

$$\bar{\boldsymbol{\lambda}}(5) = (0.12 \quad 0.16), \quad \hat{\boldsymbol{\lambda}}(5) = (0), \quad \bar{\boldsymbol{\Lambda}}(5) = (0.28).$$

The matrix $\boldsymbol{R} = (\rho_{i,j})$ of the probabilities that chain $J(t)$ will be in the state $\boldsymbol{j} \in \boldsymbol{I}(\eta)$ after leaving the set $\overline{\boldsymbol{I}}(\eta)$, is as follows:

$$\boldsymbol{R}(4) = (\boldsymbol{E} - \mathbf{diag}(\bar{\Lambda}(4))^{-1}\hat{\lambda}(4))^{-1}\mathbf{diag}(\bar{\Lambda}(4))^{-1}\bar{\lambda}(4)$$

$$= \left(\boldsymbol{E} - \begin{pmatrix} 0.14 & 0 \\ 0 & 0.20 \end{pmatrix}^{-1} \begin{pmatrix} 0 & 0.1 \\ 0 & 0 \end{pmatrix}\right)^{-1}$$

$$\times \begin{pmatrix} 0.14 & 0 \\ 0 & 0.20 \end{pmatrix}^{-1} \begin{pmatrix} 0.04 \\ 0.20 \end{pmatrix} = \begin{pmatrix} 1 \\ 1 \end{pmatrix},$$

$$\boldsymbol{R}(5) = (\mathbf{1} - (\mathbf{0.28})^{-1}(\mathbf{0}))^{-1}(\mathbf{0.28})^{-1}(\mathbf{0.12} \quad \mathbf{0.16}) = (\mathbf{0.43} \quad \mathbf{0.57}),$$

The result for $R(4)$ is self-evident, confirming that the calculations are correct.

Now, we can calculate the matrix $\tilde{\boldsymbol{\lambda}} = (\tilde{\boldsymbol{\lambda}}_{i,j})$, $\boldsymbol{i}, \boldsymbol{j} \in \boldsymbol{I}(\eta)$ of the transient intensities between the states of the set $\boldsymbol{I}(\eta)$ during a time when the considered arc e_η is operational. The sub-matrices $\boldsymbol{\lambda}^* = (\boldsymbol{\lambda}_{i,j})$ and $\boldsymbol{\lambda}^{**} = (\boldsymbol{\lambda}_{i,\nu})$, $\boldsymbol{\nu} \in \overline{\boldsymbol{I}}(\eta)$ are as follows:

$$\boldsymbol{\lambda}^*(4) = (0), \quad \boldsymbol{\lambda}^{**}(4) = (\mathbf{0.12} \quad \mathbf{0.16}),$$

$$\boldsymbol{\lambda}^*(5) = \begin{pmatrix} 0 & 0.1 \\ 0 & 0 \end{pmatrix}, \quad \boldsymbol{\lambda}^{**}(5) = \begin{pmatrix} 0.04 \\ 0.20 \end{pmatrix}.$$

With respect to formula (3.18), we have

$$\tilde{\boldsymbol{\lambda}}(4) = \boldsymbol{\lambda}^*(4) + \boldsymbol{\lambda}^{**}(4)\boldsymbol{R}(4) = (\mathbf{0.12} \quad \mathbf{0.16}) \begin{pmatrix} 1 \\ 1 \end{pmatrix} = \mathbf{0.28},$$

$$\tilde{\boldsymbol{\lambda}}(5) = \boldsymbol{\lambda}^*(5) + \boldsymbol{\lambda}^{**}(5)\boldsymbol{R}(5) = \begin{pmatrix} 0 & 0.1 \\ 0 & 0 \end{pmatrix} + \begin{pmatrix} 0.04 \\ 0.20 \end{pmatrix} (\mathbf{0.43} \quad \mathbf{0.57})$$

$$= \begin{pmatrix} 0.017 & 0.123 \\ 0.086 & 0.114 \end{pmatrix}.$$

Now, we can calculate the matrix $\Pr(\eta) = \widetilde{\boldsymbol{P}}(\boldsymbol{d}_\eta)$ of the probabilities of the final states for set $\boldsymbol{I}(\eta)$. If $\in \overline{\boldsymbol{I}}(\eta)$, then the corresponding final probability of state $\boldsymbol{j}$ equals zero. If the initial state i belongs to set $\overline{\boldsymbol{I}}(\eta)$ and the final state j belongs to set $\boldsymbol{I}(\eta)$, the nested results yield formula (3.19).

Below, the matrices $\Pr(\eta)$ are presented for some η:

$$\mathbf{Pr}(0) = \begin{pmatrix} \mathbf{0.946} & \mathbf{0.020} & \mathbf{0.033} \\ \mathbf{0.084} & \mathbf{0.820} & \mathbf{0.096} \\ \mathbf{0.015} & \mathbf{0.296} & \mathbf{0.639} \end{pmatrix},$$

$$\mathbf{Pr}(1) = \begin{pmatrix} \mathbf{0.873} & \mathbf{0.046} & \mathbf{0.081} \\ \mathbf{0.139} & \mathbf{0.693} & \mathbf{0.169} \\ \mathbf{0.026} & \mathbf{0.270} & \mathbf{0.703} \end{pmatrix},$$

$$\mathbf{Pr}(2) = \begin{pmatrix} \mathbf{0.850} & \mathbf{0.053} & \mathbf{0.097} \\ \mathbf{0.144} & \mathbf{0.677} & \mathbf{0.179} \\ \mathbf{0.172} & \mathbf{0.389} & \mathbf{0.439} \end{pmatrix},$$

$$\mathbf{Pr}(3) = \begin{pmatrix} \mathbf{0.910} & \mathbf{0.042} & \mathbf{0.048} \\ \mathbf{0.136} & \mathbf{0.735} & \mathbf{0.130} \\ \mathbf{0.047} & \mathbf{0.473} & \mathbf{0.481} \end{pmatrix},$$

$$\mathbf{Pr}(4) = \begin{pmatrix} \mathbf{0} & \mathbf{1} & \mathbf{0} \\ \mathbf{0} & \mathbf{1} & \mathbf{0} \\ \mathbf{0} & \mathbf{1} & \mathbf{0} \end{pmatrix}, \quad \mathbf{Pr}(5) = \begin{pmatrix} \mathbf{0.736} & \mathbf{0} & \mathbf{0.264} \\ \mathbf{0.456} & \mathbf{0} & \mathbf{0.544} \\ \mathbf{0.245} & \mathbf{0} & \mathbf{0.755} \end{pmatrix}.$$

Tables 3.11 and 3.12 contain the main results if the initial state is i.

As mentioned above, the minimal mean times from the source to the target are 5.643, 7.339, and 8.225 for different initial states of the random environment.

Let us now consider a stationary case, using the data of Example 3.1. Since the third eigenvalue equals to zero, then as earlier

$$\boldsymbol{\beta}_{i^*} = \boldsymbol{\beta}_2 = (\mathbf{0.577} \quad \mathbf{0.577} \quad \mathbf{0.577})^T,$$
$$\bar{\boldsymbol{\beta}}_{i^*} = \bar{\boldsymbol{\beta}}_2 = (\mathbf{0.481} \quad \mathbf{0.561} \quad \mathbf{0.690})$$

Table 3.11. Matrix of the mean shortest times $\tau_{i,\rho}$ to the target from vertex v_ρ.

i/ρ	0	1	2	3	4	5	6
0	5.643	5.275	4.275	4.294	1.570	1.260	2.300
1	7.339	7.280	7.139	5.821	2.340	2.240	4.170
2	8.225	9.438	6.517	6.542	2.480	2.790	5.010

Table 3.12. Matrix of the optimal following vertices $(u_{i,\rho})$.

i/ρ	0	1	2	3	4	5	6
0	2	3	5	6	7	7	7
1	3	3	5	4	7	7	7
2	3	3	4	4	7	7	7

Table 3.13. Mean time $\mu s(n)_\eta$ for the stationary case.

$\eta=0$	$\eta=1$	$\eta=2$	$\eta=3$	$\eta=4$	$\eta=5$	$\eta=6$	$\eta=7$	$\eta=8$	$\eta=9$	$\eta=10$	$\eta=11$
1.11	1.54	1.64	2.06	8.29	4.68	4.77	3.66	2.86	2.20	2.26	4.23

Table 3.14. The stationary mean shortest time to the target.

$\rho=0$	$\rho=1$	$\rho=2$	$\rho=3$	$\rho=4$	$\rho=5$	$\rho=6$
7.220	7.581	6.095	5.683	2.182	2.186	3.984

Therefore,

$$\lim_{t\to\infty} \boldsymbol{P}(t) = \beta_{l^*}\bar{\beta}_{i^*} = \begin{pmatrix} \mathbf{0.577} \\ \mathbf{0.577} \\ \mathbf{0.577} \end{pmatrix} (\mathbf{0.481} \quad \mathbf{0.561} \quad \mathbf{0.690})$$

$$= \begin{pmatrix} \mathbf{0.278} & \mathbf{0.324} & \mathbf{0.398} \\ \mathbf{0.278} & \mathbf{0.324} & \mathbf{0.398} \\ \mathbf{0.278} & \mathbf{0.324} & \mathbf{0.398} \end{pmatrix}.$$

Also, the stationary probabilities

$$\pi = (\pi_0, \pi_1, \pi_2) = (\mathbf{0.278} \quad \mathbf{0.324} \quad \mathbf{0.398}).$$

Table 3.13 contains the matrix of the stationary mean time $\boldsymbol{\mu s}(\boldsymbol{\eta})_i$ of the passing of arc e_η (see formula (3.23)).

The vector of the stationary mean shortest time τ_ρ to the target from vertex v_ρ is presented in Table 3.14.

In addition, the minimum stationary mean time from the source to the target equals 7.220. It is interesting to compare this with the case in which information about the state of the random environment is not used. In this case, we are guided by the mean time $\mu s(\eta)$ of overcoming arc e_η. The corresponding values are presented in Table 3.13. It is obvious that the shortest path passes through vertices v_0, v_3, v_4, v_7 or arcs e_2, e_7, e_9. Table 3.13 shows that this path has the following length:

$$1.64 + 3.66 + 2.20 = 7.50.$$

We see that the result is worse on $\frac{7.50-7.22}{7.22}100\% = 3.73\%$.

3.9 Concluding remarks

The classical problem of shortest paths in a finite network is considered when the network operates in a random environment. The last is presented by a continuous-time finite Markov chain $J(t)$. The main results were obtained by substituting the argument time t with passed distance x along the current link.

Further research will focus on the search for the path with the maximal probability of transportation for a given time.

Appendix

Here, we discuss the problem of choosing Δ-value that is used in the *delta-method.*

Let us hold some arc by number η with speed $s_{i,\eta} > 0$ in state i. Relation $\Delta/s_{i,\eta}$ gives the time of covering the distance Δ. Value $\Lambda_i\Delta/s_{i,\eta}+o(\Delta)$ is a probability that the random environment outputs from state i during this time. The delta-method is based on the condition that the value Δ is so small, that $o(\Delta)$ can be rejected. We wish to give an upper limit of mistake probability.

Let $\alpha = \max\{\Lambda_i/s_{i,\eta} : i = 1, \ldots, k\}$. The probability of mistakes grows if for all states $i = 1, \ldots, k$ to change $\Lambda_i/s_{i,\eta}$ by α. Now we have a Poisson process with an intensity α. Let $A(1)$ be a random event: two or more events of this process occur during time Δ. The probability of this event is as follows:

$$\Pr\{A(1)\} = 1 - \exp(-\alpha\Delta) - \alpha\Delta\exp(-\alpha\Delta).$$

Further, let the length of arc $l_\eta = n\Delta$. Considering a random event $A(n)$, we assume that event $A(1)$ occurs during n observations. This is the sum of n events $A(1)$.

It is necessary to choose Δ-value such that the probability $\Pr\{A(n)\}$ will be less than the given small $\xi > 0$. This probability is less than the sum of the probabilities $\Pr\{A(1)\}$. Therefore, we obtain the following inequality:

$$\frac{l_\eta}{\Delta}[1 - \exp(-\alpha\Delta) - \alpha\Delta\exp(-\alpha\Delta)] \le \xi. \tag{A.1}$$

The maximal value of Δ, that satisfies this inequality is found numerically by enumeration.

The final value of Δ is determined as the maximal one for all arcs.

Let us return to the considered example and set $\xi = 0.01$. The "worst" arc is the sixth arc. It has the following characteristics: $\alpha = \max\{\Lambda_6/s_{i,6} : i = 1, \ldots, k\} = 0.07$, $l_6 = 11$.

An enumeration with step 0.0025 yielded 0.038 as the maximal value of Δ. Note that this result was obtained for the worst case, the value $\Delta = 0.05$ has been used in the previous calculations.

References

[1] Christofides, N., *Graph Theory. Algorithmic Approach.* Academic Press, New York, London, San Francisco, 1975.

[2] Hu, T.C., *Integer Programming and Network Flow.* Addison-Wesley, California, London, Don Mills, Ontario, 1970.

[3] Azaron, A., Kianfar, F., Dynamic shortest path in stochastic dynamic networks: Ship routing problem. *European Journal of Operational Research*, 2003;144(1):138–156.

[4] Azaron, A., Modareres, M., Distribution function of the shortest path in networks of queues. *OR Spectrum*, 2005;27:123–144.

[5] Bertsekas, D.P., Titsiklis, J.N., An analysis of stochastic shortest path problems. *Mathematics of Operations Research*, 1991;16(3):580–595.

[6] Fan, Y.Y., Kalaba, R.E., Moore J.E., Shortest paths in stochastic networks with correlated link costs. *Computer & Mathematics with Applications*, 2005;49:1649–1564.

[7] Hutson, K.R., Shier, D.R., Extended dominance and a stochastic shortest path problem. *Computer & Operations Research*, 2009;36(2):584–596.

[8] Ji, X., Models and algorithm for stochastic shortest path problem. *Applied Mathematics and Computation*, 2005;170(1):503–514.

[9] Kijima, M., *Markov Processes for Stochastic Modelling.* The University Press, Cambridge, UK, 1997.

[10] Pacheco, A., Tang, L.C., Prabhu, N.U., *Markov-Modulated Processes & Semiregenerative Phenomena.* World Scientific, New Jersey, London, 2009.

[11] Bellman, R., *Introduction to Matrix Analysis.* McGraw-Hill Book Company, Inc., New York, Toronto, London, 1960.

[12] Soyer, Markov and hidden Markov models. In: James J. Cochran (ed.), *Wiley Encyclopedia of Operation Research and Management Sciences*, John Wiley & Sons, Inc., New York, 2010.

Chapter 4

Network Flows with Markov-Modulated Costs of Arcs

4.1 Positing the problem

In this chapter, we consider *network* $N = (V, E)$ [1–6] with a set $V = (v_0, v_1, \ldots, v_m)$ of the *vertices* and a set $E = (e_1, e_2, \ldots, e_n)$ of the directed *arcs*. Each arc is of the form $e = (v_w, v_u)$, where $w < u$. Two vertices are distinguished: *source* v_0 and *target* v_n. Let S_v^+ and S_v^- be the set of arcs, which have vertex $v \in V$ as their own input and output, respectively.

As usual, the flow in network N is defined as a real function f on E, subject to the following constraints:

$$f(e) \geq 0, \quad e \in E,$$
$$\sum_{e \in S_v^+} f(e) = \sum_{e \in S_v^-} f(e), \quad v \in V - \{s, t\}. \tag{4.1}$$

The value of the flow f is defined as follows:

$$val(f) = \sum_{e \in S_0^-} f(e) = \sum_{e \in S_m^+} f(e). \tag{4.2}$$

The network operates in a random environment described by an irreducible continuous-time Markov chain $J(t), t > 0$ [7,8]. This chain has a finite number k of states. The transition intensities between the states are given by the matrix $\lambda = (\lambda_{i,j})$.

Each arc e has a cost per unit of flow transfer. This cost depends on the state of the random environment $J(t)$ and equals $c_i(e)$ if $J(t) = i$, $i \in \{1, \dots, k\}$.

Our goal is to compute the Laplace transform of the flow cost $C(f, t)$ in the time interval $(0, t)$ for stationary and non-stationary cases, given flow f.

The remainder of this chapter is organized as follows. The main analytical results for the random environment are presented in Section 4.2. The stationary case is considered in Section 4.3. The Laplace transform of the flow cost for the non-stationary case is derived in Section 4.4. The mean cost of flow, together with the probability of a final state, is considered in Section 4.5. Section 4.6 contains a numerical example. The chapter ends with a conclusion.

4.2 A distribution of the states of the random environment

Let $\Lambda_i = \sum_{\varsigma=1}^{k} \lambda_{i,\varsigma}$, $\Lambda = (\Lambda_1, \dots, \Lambda_k)^T$. The $k \times k$-matrix $A = \lambda - \mathrm{diag}(\Lambda)$ is called *the generator* of the Markov chain. We denote the eigenvalues and eigenvectors of the matrix as $\chi_1, \chi_2, \dots, \chi_k$ and $\beta_1, \beta_2, \dots, \beta_k$, respectively. It is supposed that all values $\chi_1, \chi_2, \dots, \chi_k$ are different, one eigenvalue equals zero, and others are negative. We assume that $\chi_1 = 0$.

The matrix of transition probabilities during time t is calculated as follows:

$$P(t) = (P_{i,j}(t))_{\boldsymbol{k}\times\boldsymbol{k}} = \sum_{\boldsymbol{i=1}}^{\boldsymbol{k}} \exp(\chi_i t)\beta_i\tilde{\beta}_i$$

$$= B\,\mathrm{diag}(\exp(\chi t))B^{-1}, \quad t \geq 0, \tag{4.3}$$

where $\exp(\chi t) = (\exp(\chi_i t) \ \cdots \ \exp(\chi_k t))^{\mathrm{T}}$, $B^{-1} = (\tilde{\beta}_1^T \ \tilde{\beta}_2^T \ \tilde{\beta}_3^T)^T$.

The expectation of the sojourn time $T_{i,j}(t)$ in state j on interval $(0, t)$, if the initial state equals i, is calculated for $i, j \in \{1, \dots, k\}$ as follows:

$$E(T_{i,j}(t)) = \int_0^t P_{i,j}(u)du = \int_0^t \sum_{\eta=1}^{k} B_{i,\eta} \exp(\chi_\eta u) B_{\eta,j}^{-1} du$$

$$= tB_{i,1}B_{1,j}^{-1} + \sum_{\eta=2}^{k} B_{i,\eta} \int_0^t \exp(\chi_\eta u) du B_{\eta,j}^{-1}$$

$$= tB_{i,1}B_{1,j}^{-1} + \sum_{\eta=2}^{k} B_{i,\eta} \frac{1}{\chi_\eta} (\exp(\chi_\eta t) - 1) B_{\eta,j}^{-1}. \quad (4.4)$$

Additionally, the eigenvalue χ_1 is equal to zero, whereas the other values are negative. All components of the eigenvector corresponding to the zero eigenvalue are the same. Then,

$$\lim_{t\to\infty} P(t) = \lim_{t\to\infty} \sum_{i=1}^{k} \exp(\chi_i t)\beta_i \tilde{\beta}_i = \beta_1 \tilde{\beta}_1. \quad (4.5)$$

Because all components of vector β_1 are the same, all the rows of matrix $\beta_1\tilde{\beta}_1$ are the same. Each row represents a vector of stationary probabilities of states $p = (p_1, \ldots, p_k)$.

4.3 A mean cost of the flow for stationary case

Let $C(f)$ be the full flow cost per unit time for the stationary case. The stationary mean is calculated as follows:

$$E(C(f)) = \sum_{e\in E} f(e) \sum_{i=1}^{k} p_i c_i(e). \quad (4.6)$$

Let all arcs be enumerated, with $|e|$ being the ordinal number of arc e and $\eta = |e|$ being the total notation of arc's ordinal numbers, $\eta = 1, \ldots, n$. Furthermore, let $f_\eta = f(e)$, $c_{i,\eta} = c_i(e)$, $f = (f_1, \ldots, f_n)^T$, $c = (c_{i,\eta})_{k\times n}$. Thus, Eq. (4.6) can be transformed as follows:

$$E(C(f)) = \sum_{\eta=1}^{n} f_\eta \sum_{i=1}^{k} p_i c_{i,\eta} = pcf. \quad (4.7)$$

Analogously, let all vertices be enumerated, with $|v|$ being the ordinal number of vertex v and $\psi = |v|$ being the total notation of the vertex's ordinal numbers, $\psi = 0, \ldots, m$. We suppose that if

$e = (v', v'')$, then $|v'| < |v''|$. Furthermore, let $\widetilde{S}_\psi^+$ and $\widetilde{S}_\psi^-$ be the set of arc numbers, which have the vertex of number ψ as their own input and output, respectively.

As a result, system (4.1) can be transformed as follows:

$$f_\eta \geq 0, \quad \eta = 1, \ldots, n, \tag{4.8}$$

$$\sum_{\eta \in \widetilde{S}_\psi^+} f_\eta = \sum_{\eta \in \widetilde{S}_\psi^-} f_\eta, \quad \psi = 1, \ldots, m-1. \tag{4.9}$$

The value val^* of flow f is calculated as follows (instead of formula (4.2)):

$$val* = \sum_{\eta \in \widetilde{S}_1^-} f_\eta = \sum_{\eta \in \widetilde{S}_m^+} f_\eta. \tag{4.10}$$

4.4 Non-stationary case

Let the initial state i of the random environment be fixed, and the cost $C(f, t \,|\, i)$ of flow f during time t be interesting for us. It is assumed that the flow value f_η on the ηth arc is transferred during unit time. If $T_{i,j}(t)$ is the sojourn time in state j on the interval $(0, t)$, then,

$$C(f, t \,|\, i) = \sum_{\eta=1}^{n} f_\eta \sum_{j=1}^{k} T_{i,j}(t) c_{j,\eta}, \quad t \geq 0.$$

The mean value is calculated simply as follows:

$$E(C(f, t \,|\, i)) = \sum_{\eta=1}^{n} f_\eta \sum_{j=1}^{k} E(T_{i,j}(t)) c_{j,\eta}. \tag{4.11}$$

Now, we need to calculate the Laplace transform [9] of the cost $C(f, t)$ during time t. Let $\varphi_{i,j}(s, t)$ be this transform jointly with the probability that state j takes place at time t, if the initial state equals i:

$$\varphi_{i,j}(s, t) = E(\exp(-sC(f, t))\delta(J(t) = j) \,|\, J(0) = i), \tag{4.12}$$

where $\delta(J(t) = j)$ is the indicator of the event $\{J(t) = j\} : \delta(J(t) = j) = 1$ if this event occurs and 0 otherwise.

By applying the usual reasoning, we obtain

$$\begin{aligned}
\varphi(s,t+\Delta t)_{i,j} &= (1-\Lambda_i\Delta t)e^{-s\Delta t\sum_{\eta=1}^{n} f_\eta c_{i,\eta}}\varphi(s,t)_{i,j} \\
&\quad + \Delta t\sum_{\xi=1}^{k}\lambda_{i,\xi}\varphi(s,t)_{\xi,j} + o(\Delta t) \\
&= \varphi(s,t)_{i,j}(1-\Lambda_i\Delta t)(1-s\Delta t\sum_{\eta=1}^{n} f_\eta c_{i,\eta}) \\
&\quad + \Delta t\sum_{\xi=1}^{k}\lambda_{i,\xi}\varphi(s,t)_{\xi,j} + o(\Delta t),
\end{aligned}$$

$$\frac{\partial}{\partial t}\varphi(s,t)_{i,j} = \varphi(s,t)_{i,j}\left(-\Lambda_i - s\sum_{\eta=1}^{n} f_\eta c_{i,\eta}\right) + \sum_{\xi=1}^{k}\lambda_{i,\xi}\varphi(s,t)_{\xi,j}$$

The matrix representation is the following:

$$\frac{\partial}{\partial t}\varphi(s,t) = -D\varphi(s,t) + \lambda\varphi(s,t) = (\lambda - D)\varphi(s,t), \tag{4.13}$$

where D is the diagonal matrix with the diagonal elements

$$D_{i,i} = \left(\Lambda_i + s\sum_{\eta=1}^{n} f_\eta c_{i,\eta}\right).$$

If $c = (c_{i,\eta})$ is $k\times n$-matrix, $f = (f_1,\dots f_n)^T$ is n-vector, then

$$D = \mathrm{diag}(\Lambda + scf). \tag{4.14}$$

The solution of system (4.13) can be represented by the matrix exponent:

$$\begin{aligned}
\varphi(s,t) &= \exp(t(\lambda - D)) = I + \sum_{\xi=1}^{\infty}\frac{1}{\xi!}(\lambda - D)^\xi t^\xi \\
&= I + \sum_{\xi=1}^{\infty}\frac{1}{\xi!}(\lambda - \mathrm{diag}(\Lambda + scf))^\xi t^\xi, \quad t \ge 0. \qquad (4.15)
\end{aligned}$$

Note that if $s = 0$, then $\varphi(0,t) = P(t)$, and formula (4.13) gives the true result:

$$\frac{\partial}{\partial t}P(t) = (\lambda - \mathrm{diag}(\Lambda))P(t).$$

If the last state of the random environment is not of interest to us, then we have the vector $\hat{\varphi}(s,t) = \varphi(s,t)\Delta$ instead of the matrix $\varphi(s,t)$:

$$\hat{\varphi}(s,t) = \varphi(s,t)\Delta = \Delta - \sum_{\xi=1}^{\infty}\frac{t^{\xi}}{\xi!}(\lambda - D)^{\xi-1}scf,\, s,\, t \geq 0, \tag{4.16}$$

where Δ is k-dimensional unit vector.

4.5 Mean cost of flow jointly with the probability of a final state for the non-stationary case

Let $\mu(f,t)_{i,j}$ be the mean cost of the flow jointly with the probability that state j takes place at time t, if the initial state equals i. The corresponding matrix $\boldsymbol{\mu}(f,t) = (\mu(f,t)_{i,j})$ is calculated as follows:

$$\boldsymbol{\mu}(f,t) = -\frac{\partial}{\partial s}\varphi(s,t)|_{s=0} = -\sum_{\xi=1}^{\infty}\frac{1}{\xi!}t^{\xi}\frac{\partial}{\partial s}(\lambda - D)^{\xi}|_{s=0}. \tag{4.17}$$

Derivatives in the last equation are calculated as follows:

$$\frac{\partial}{\partial s}(\lambda - D)^{\xi} = -\sum_{\varsigma=0}^{\xi-1}(\lambda - D)^{\varsigma}\,\mathrm{diag}(cf)(\lambda - D)^{\xi-\varsigma-1}, \tag{4.18}$$

where $(\lambda - D)^0$ is the unit matrix.

As $\lambda - D = \lambda - \mathrm{diag}(\Lambda + scf)$, then

$$-\frac{\partial}{\partial s}(\lambda - D)^{\xi}|_{s=0} = \sum_{\varsigma=0}^{\xi-1}(\lambda - \mathrm{diag}(\Lambda))^{\varsigma}\,\mathrm{diag}(cf)(\lambda - \mathrm{diag}(\Lambda))^{\xi-\varsigma-1}.$$

Also,

$$\begin{aligned}\boldsymbol{\mu}(f,t) &= -\frac{\partial}{\partial s}\varphi(s,t)|_{s=0}\\ &= t\,\mathrm{diag}(cf) + \sum_{\xi=2}^{\infty}\frac{1}{\xi!}t^{\xi}\sum_{\varsigma=0}^{\xi-1}(\lambda - \mathrm{diag}(\Lambda))^{\varsigma}\,\mathrm{diag}(cf)\\ &\quad\times(\lambda - \mathrm{diag}(\Lambda))^{\xi-\varsigma-1}. \end{aligned} \tag{4.19}$$

If the last state of the random environment is not of interest to us, then we have the vector $\hat{\boldsymbol{\mu}}(f,t) = \boldsymbol{\mu}(f,t)\Delta$ instead of the matrix $\boldsymbol{\mu}(f,t)$. As a result, expression (4.19) has the form:

$$\begin{aligned}\hat{\boldsymbol{\mu}}(f,t) &= \boldsymbol{\mu}(f,t)\Delta\\ &= tcf + \sum_{\xi=2}^{\infty}\frac{1}{\xi!}t^{\xi}\sum_{\varsigma=0}^{\xi-2}(\lambda - \mathrm{diag}(\Lambda))^{\varsigma}\,\mathrm{diag}(cf)\\ &\quad\times(\lambda - \mathrm{diag}(\Lambda))^{\xi-\varsigma-2}cf, \quad t \geq 0. \end{aligned} \tag{4.20}$$

The following identity allows verification of the calculation:

$$\hat{\boldsymbol{\mu}}(f,t) = \begin{pmatrix} E(C(f,t\,|\,1)) \\ \cdots \\ E(C(f,t\,|\,k)) \end{pmatrix}. \tag{4.21}$$

It is interesting to consider the specific cost of the flow in the interval $(0, t)$. Let us denote this using an asterisk * as

$$C^*(f,t) = \frac{1}{t}C(f,t), \quad \boldsymbol{\mu}^*(f,t) = \frac{1}{t}\boldsymbol{\mu}(f,t), \quad \hat{\boldsymbol{\mu}}^*(f,t) = \frac{1}{t}\hat{\boldsymbol{\mu}}(f,t). \tag{4.22}$$

The corresponding Laplace transform is calculated simply as follows:

$$\begin{aligned}\varphi*_{i,j}(s,t) &= E(\exp(-sC^*(f,t))\delta(J(t)=j)\,|\,J(0)=i)\\ &= E\left(\exp\left(-s\frac{1}{t}C(f,t)\right)\delta(J(t)=j)\,|\,J(0)=i\right) = \varphi_{i,j}\left(\frac{s}{t},t\right). \end{aligned}$$

Therefore,

$$\varphi*(s,t) = \varphi\left(\frac{s}{t},t\right), \quad \hat{\varphi}^*(s,t) = \hat{\varphi}\left(\frac{s}{t},t\right). \tag{4.23}$$

4.6 Numerical example

The continuous-time Markov chain $J(t)$ has three states ($k = 3$) and the following matrix of transition intensities between the states:

$$\lambda = \begin{pmatrix} 0 & 0.1 & 0.2 \\ 0.2 & 0 & 0.3 \\ 0.3 & 0.4 & 0 \end{pmatrix}.$$

The vector $\Lambda = (\Lambda_1, \ldots, \Lambda_k)^T$ and the generator A is presented as follows:

$$\Lambda = (0.3, 0.5, 0.7)^T, \quad A = \begin{pmatrix} -0.3 & 0.1 & 0.2 \\ 0.2 & -0.5 & 0.3 \\ 0.3 & 0.4 & -0.7 \end{pmatrix}.$$

The vectors of the eigenvalues χ, matrix B of the eigenvectors, and inverse matrix B^{-1} are as follows:

$$\chi = (\chi_1 \quad \chi_2 \quad \chi_3)^T = (0 \quad -0.521 \quad -0.979)^T,$$

$$B = (\beta_1 \quad \beta_2 \quad \beta) = \begin{pmatrix} 0.577 & -0.647 & -0.187 \\ 0.577 & 0.658 & -0.464 \\ 0.577 & 0.386 & 0.866 \end{pmatrix},$$

$$B^{-1} = \left(\tilde{\beta}_1^T \quad \tilde{\beta}_2^T \quad \tilde{\beta}_3^T\right)^T = \begin{pmatrix} 0.781 & 0.509 & 0.442 \\ -0.801 & 0.634 & 0.167 \\ -0.164 & -0.622 & 0.786 \end{pmatrix}.$$

The matrix $P(t)$ of transition probabilities and the matrix $ET(t) = (ET_{i,j}(t))$, where $ET_{i,j}(t)$ is the expectation of the sojourn time in state j at interval $(0, t)$ for the initial state i, is calculated using formulas (4.3) and (4.4). The calculations yielded the following results for $t = 5$ and $t = 10$:

$$P(5) = \begin{pmatrix} 0.490 & 0.265 & 0.246 \\ 0.413 & 0.327 & 0.260 \\ 0.427 & 0.308 & 0.265 \end{pmatrix}, \quad P(10) = \begin{pmatrix} 0.454 & 0.292 & 0.254 \\ 0.448 & 0.296 & 0.255 \\ 0.449 & 0.295 & 0.255 \end{pmatrix},$$

$$ET(5) = \begin{pmatrix} 3.208 & 0.859 & 0.933 \\ 1.395 & 2.505 & 1.100 \\ 1.562 & 1.359 & 2.079 \end{pmatrix}, \quad ET(10) = \begin{pmatrix} 5.531 & 2.277 & 2.193 \\ 3.581 & 4.033 & 2.386 \\ 3.775 & 2.858 & 3.367 \end{pmatrix}.$$

The stationary probabilities of the states of the random environment are calculated using formula (4.5):

$$p = (p_1 \quad p_2 \quad p_3) = (0.451 \quad 0.294 \quad 0.255).$$

The network $N = (V, Ar)$ is represented by the matrix of its own arcs.

$$Ar = \begin{pmatrix} 1 & 2 & 3 & 4 & 5 & 6 & 7 & 8 & 9 & 10 & 11 & 12 \\ 0 & 0 & 0 & 1 & 1 & 2 & 2 & 3 & 3 & 4 & 5 & 6 \\ 1 & 2 & 3 & 3 & 6 & 4 & 5 & 4 & 6 & 7 & 7 & 7 \end{pmatrix},$$

where the columns correspond to various arcs with numbers $\eta = 1, \ldots, 12$, the first row contains the number of arcs, the second row contains the initial vertex of the arc, and the third row contains the final vertex of the arc.

We suppose that the following flow f is given:

$$\begin{pmatrix} \eta \\ f_\eta \end{pmatrix} = \begin{pmatrix} 1 & 2 & 3 & 4 & 5 & 6 & 7 & 8 & 9 & 10 & 11 & 12 \\ 4 & 5 & 6 & 2 & 2 & 3 & 2 & 2 & 6 & 5 & 2 & 8 \end{pmatrix}$$

Arc costs $\{c_{i,\eta}\}$ are yielded by the matrix

$$c = (c_{i,\eta}) = \begin{pmatrix} 1 & 2 & 3 & 1 & 1 & 3 & 2 & 3 & 2 & 3 & 2 & 3 \\ 2 & 3 & 4 & 2 & 3 & 4 & 5 & 4 & 4 & 5 & 4 & 7 \\ 5 & 4 & 5 & 6 & 3 & 4 & 6 & 5 & 6 & 5 & 5 & 8 \end{pmatrix},$$

where columns correspond to various arcs with numbers $\eta = 1, \ldots, 12$, and rows correspond to the states of the random environment with numbers $i = 1, 2, 3$.

The stationary mean flow cost per unit of time is $E(C(f)) = 172.94$ (see formula (4.7)). These data allow us to verify our results in the future.

We now consider a transient case. The initial state i of the random environment is fixed, and the cost $C(f, t \mid i)$ of the flow at time t is considered. The average cost $E(C(f, t \mid i))$ is calculated using formula (4.11). The corresponding vectors $E(C(f, t)) = (E(C(f, t \mid 1)), E(C(f, t \mid 2)), E(C(f, t \mid 3)))$ for $t = 5$ and $t = 10$ are as follows:

$$E(C(f, 5)) = (764.5 \quad 937.2 \quad 978.0),$$
$$E(C(f, 10)) = (1627 \quad 1814 \quad 1852). \tag{4.24}$$

The average specific costs during time t and various initial states i of the random environment are calculated as follows: $C^*(f, t \mid i) = \frac{1}{t} C(f, t \mid i)$. Let $E(C^*(f, t)) = (E(C^*(f, t \mid 1)), E(C^*(f, t \mid 2)), E(C^*(f, t \mid 3)))$. We obtained the following result for the large time $t = 1000$: $E(C^*(f, 100000)) = (172.94 \quad 172.94 \quad 172.94)$, which coincides with the corresponding result for the stationary case: $E(C(f)) = 172.94$.

The results are calculated using formulas (4.7) and (4.11). Let us present the results obtained using Laplace transforms. The Laplace transforms $\varphi(s, t)$ and $\hat{\varphi}(s, t) = \varphi(s, t)\Delta$ are calculated using formulas (4.15) and (4.16), where the infinite sum is replaced by a finite sum with number $n\max$ of summands. Some values for $n\max = 70$ are as follows:

$$\varphi(0.005, 5) = \begin{pmatrix} 0.022 & 5.040 \times 10^{-3} & 4.617 \times 10^{-3} \\ 8.395 \times 10^{-3} & 2.755 \times 10^{-3} & 2.112 \times 10^{-3} \\ 7.767 \times 10^{-3} & 2.25410^{-3} & 1.860 \times 10^{-3} \end{pmatrix},$$

$$\varphi(0.005, 10) = \begin{pmatrix} 5.428 \times 10^{-4} & 1.329 \times 10^{-4} & 1.187 \times 10^{-4} \\ 2.205 \times 10^{-4} & 5.466 \times 10^{-5} & 4.845 \times 10^{-5} \\ 2.008 \times 10^{-4} & 4.945 \times 10^{-5} & 4.421 \times 10^{-5} \end{pmatrix},$$

$$\hat{\varphi}(0.005, 5) = (0.031 \quad 0.013 \quad 0.012),$$

$$\hat{\varphi}(0.005, 10) = (7.945 \times 10^{-4} \quad 3.236 \times 10^{-4} \quad 2.945 \times 10^{-4}).$$

Figure 4.1 presents graphics of functions $\varphi t(s, t, n\max)_{i-1} = \hat{\varphi}(s, t)_i$ for $t = 5$ and $n\max = 70$.

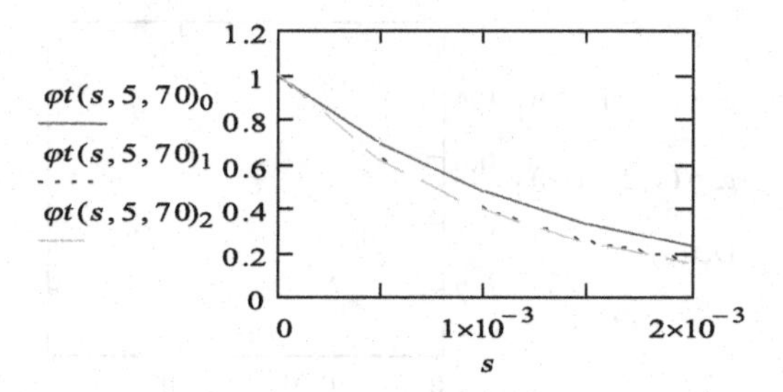

Fig. 4.1. Graphics of Laplace transforms $\hat{\varphi}(s,t)_i$.

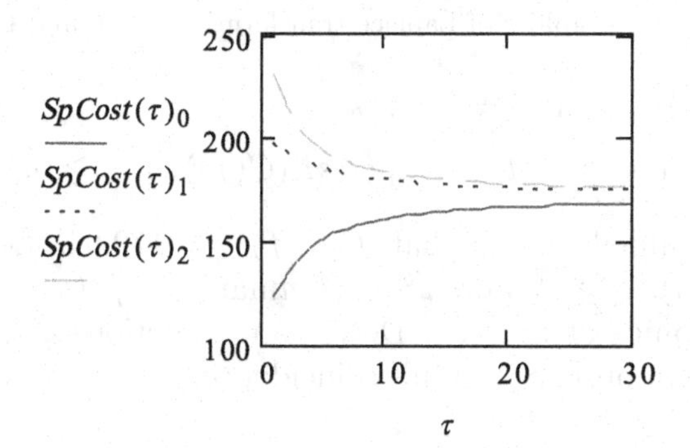

Fig. 4.2. Graphics of averages $E(\hat{C}^*(f,t\,|\,i))$.

The matrices $\boldsymbol{\mu}(f,t) = (\mu(f,t)_{i,j})$ of the mean costs jointly with the probability of the final state are as follows for $n\text{max} = 100$:

$$\boldsymbol{\mu}(f,5) = \begin{pmatrix} 326.50 & 224.84 & 213.18 \\ 344.47 & 325.59 & 267.11 \\ 372.38 & 321.08 & 284.50 \end{pmatrix},$$

$$\boldsymbol{\mu}(f,10) = \begin{pmatrix} 686.18 & 500.81 & 440.21 \\ 761.36 & 562.84 & 489.54 \\ 780.45 & 572.62 & 499.12 \end{pmatrix}.$$

The sum of their columns yields the previous results (4.24) exactly (see formulas (4.20) and (4.21)).

The dependence of the specific averages $SpCost(\tau)_{i-1} = E(C^*(f,t\,|\,i))$ on t and i, which are calculated by formula (4.22), is presented in Fig. 4.2.

Using the Laplace transform $\hat{\varphi}^*(s,t)_i$ of the specific cost of the flow on the interval $(0,t)$, we can verify our results for the stationary

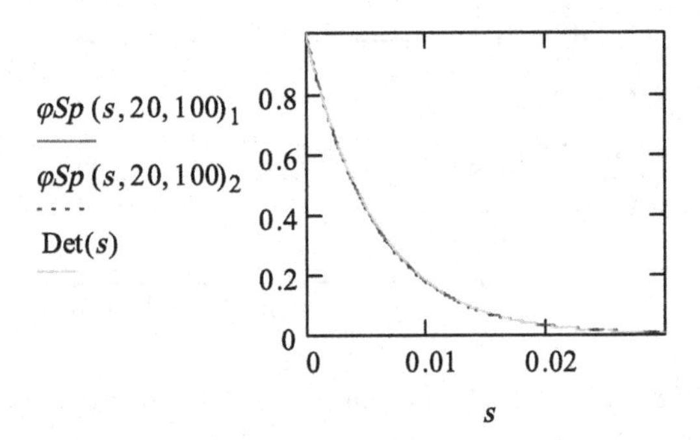

Fig. 4.3. Graphics of Laplace transforms $\hat{\varphi}^*(s,t)_i$ and $\text{Det}(s)$.

case. Evidently, for all $i = 1, \ldots, k$

$$\lim_{t\to\infty} \hat{\varphi}^*(s,t)_i = \exp(-sE(C(f))), \quad s \geq 0.$$

We determined earlier that $E(C(f)) = 172.94$. Let $\text{Det}(s) = \exp(-172.94s)$, $s \geq 0$, and $\varphi\text{Sp}(s,t,nmax)_i = \hat{\varphi}^*(s,t)_i$. Figure 4.3 presents graphics of function $\text{Det}(s)$ and functions $\varphi\text{Sp}(s,20,100)_i$ for $i = 1, 2$. We observe the full coincidence.

4.7 Optimization of the stationary flow

In the preceding analysis, we considered the network $N = (V, E)$ with a set $V = (v_0, v_1, \ldots, v_m)$ of the vertices and a set $E = (e_1, e_2, \ldots, e_n)$ of the directed arcs. A flow $f(e) \geq 0$, $e \in E$, is determined using formulas (4.1). A value of the flow $val(f)$ is determined by employing formula (4.2). In relation to formula (4.6), the stationary cost $\tilde{c}(e)$ per unit flow transfer on the arc e is calculated as follows:

$$\tilde{c}(e) = \sum_{i=1}^{k} p_i c_i(e). \tag{4.25}$$

A cost $C(f)$ of the full flow per unit time has the following stationary mean:

$$E(C(f)) = \sum_{e\in E} f(e)\tilde{c}(e). \tag{4.26}$$

Now, we aim to decrease this cost. In order to achieve that, it is necessary to determine arc capacities $\beta(e) \geq f(e)$, $e \in E$. If, for all arcs, $\beta(e) = f(e)$, then the flow f is the unique flow with the value $val(f)$. Therefore, this flow cannot be improved. We further assume that there exist arcs with a strong inequality $\beta(e) > f(e)$.

Also, our objective is to find an admissible flow f^* with the value $val(f)$ and a minimal mean cost $E(C(f*))$. We will now present the procedures below that solve this problem.

(1) *Incremental graph* (*IG*)

This graph has the same set of vertices V. The arc $e \in E$ belongs to the graph *IG* if $f(e) < \beta(e)$. Such an arc is called *a direct arc* and is denoted $\vec{e}$. If $f(e) > 0$ then this arc is considered *an inverse arc* $\overleftarrow{e}$ in the graph *IG* and is denoted as $\overleftarrow{e}$.

The following costs are given to arcs $\vec{e}$ and $\overleftarrow{e}$:

$$c(\vec{e}) = c(e), \quad c(\overleftarrow{e}) = -c(e). \tag{4.27}$$

(2) *Contours of the incremental graph*

A contour, denoted as *Co*, is a sequence of different arcs $e(1), e(2), \ldots, e(\theta)$, satisfying the following conditions: if $e(\eta) = (v' \;\; \mathrm{v}'')^T$, then $e(\eta + 1) = (v'' \;\; \mathrm{v}''')^T$ for $\eta = 1, \ldots, \theta - 2$, and the input of arc $e(\theta)$ and output of arc $e(1)$ coincide.

We will consider the contours of *Incremental graph* (*IG*). The arcs of this graph have costs with respect to formula (4.27). The cost of the contour *Co* is the sum of all its arcs:

$$c(Co) = \sum_{e \in Co} c(e). \tag{4.28}$$

There are known algorithms for enumerating all contours of a directed graph. If there is no contour with a negative cost, the current flow is considered optimal [1,3]. However, if such a contour exists, then a transfer to the following procedure occurs.

(3) *Decrease of flow costs*

Each arc $e(\eta)$ of the contour *Co* gets a possible value $d(\eta)$ of flow's change:

$$d(\eta) = \begin{bmatrix} \beta(e(\eta)) - f(e(\eta)) & \text{if } e(\eta) \text{ is a direct arc,} \\ f(e(\eta)) & \text{if } e(\eta) \text{ is an inverse arc.} \end{bmatrix} \tag{4.29}$$

The minimal value, denoted as d^*, in the sequence $d(1), d(2), \ldots, d(\theta)$ allows for a modification of the flow in this way.

Each arc $e(\eta)$ of the contour Co receives a new flow value:

$$f(e(\eta)) = \begin{bmatrix} f(e(\eta)) + d^*, & \text{if } e(\eta) \text{ is a direct arc,} \\ f(e(\eta)) - d^*, & \text{if } e(\eta) \text{ is an inverse arc.} \end{bmatrix} \tag{4.30}$$

The cost of the modified flow is less on $c(Co)d^*$.

(4) *A transfer on the procedure*

We will refer to the complete or partial execution of the described procedures as an *iteration*. If all procedures are performed, a transfer to new iteration occurs.

Continuing with our example, the first step is to determine the capacities of the given arcs $\{\beta(e)\}$. Let the capacity of each arc be one more than its flow value:

$$\begin{pmatrix} \eta \\ \beta(\eta) \end{pmatrix} = \begin{pmatrix} 1 & 2 & 3 & 4 & 5 & 6 & 7 & 8 & 9 & 10 & 11 & 12 \\ 5 & 6 & 7 & 3 & 3 & 4 & 3 & 3 & 7 & 6 & 3 & 9 \end{pmatrix}.$$

We consider a stationary flow. The stationary probabilities of the random environment's states are known:

$$p = (p_1 \quad p_2 \quad p_3) = (0.451 \quad 0.294 \quad 0.255).$$

Therefore stationary mean costs per unit time for different arcs are the following (see formula (4.25)):

$$\begin{aligned} &\tilde{c}(e(1), \ldots, e(12)) \\ &= pc = (p_1 \quad p_2 \quad p_3)(c_{i,\eta}) \\ &= (0.451 \quad 0.294 \quad 0.255) \begin{pmatrix} 1 & 2 & 3 & 1 & 1 & 3 & 2 & 3 & 2 & 3 & 2 & 3 \\ 2 & 3 & 4 & 2 & 3 & 4 & 5 & 4 & 4 & 5 & 4 & 7 \\ 5 & 4 & 5 & 6 & 3 & 4 & 6 & 5 & 6 & 5 & 5 & 8 \end{pmatrix} \\ &= (2.31 \;\; 2.80 \;\; 3.80 \;\; 2.57 \;\; 2.10 \;\; 3.55 \;\; 3.90 \;\; 3.80 \;\; 3.61 \;\; 4.10 \;\; 3.35 \;\; 5.45). \end{aligned}$$

We wish to find an admissible flow f^* of the value $val(f^*) = 15$ with minimal mean cost $E(C(f^*))$, see formulas (4.6) and (4.7), using the above-described procedures.

We have the starting flow as follows:

$$f = (4 \quad 5 \quad 6 \quad 2 \quad 2 \quad 3 \quad 2 \quad 2 \quad 6 \quad 5 \quad 2 \quad 8).$$

Corresponding incremental graph is the following (the big number 999 means that an arc is absent):

$$\text{IG:=} \begin{pmatrix} 999 & 2.314 & 2.804 & 3.804 & 999 & 999 & 999 & 999 \\ -2.314 & 999 & 999 & 2.569 & 999 & 999 & 2.098 & 999 \\ -2.804 & 999 & 999 & 999 & 3.549 & 3.902 & 999 & 999 \\ -3.804 & -2.569 & 999 & 999 & 3.804 & 999 & 3.608 & 999 \\ 999 & 999 & -3.549 & -3.804 & 999 & 999 & 999 & 4.098 \\ 999 & 999 & -3.902 & 999 & 999 & 999 & 999 & 3.353 \\ 999 & -2.098 & 999 & -3.608 & 999 & 999 & 999 & 5.451 \\ 999 & 999 & 999 & 999 & -4.098 & -3.353 & -5.451 & 999 \end{pmatrix}$$

The first considered contour with a negative cost is the following:

$$Co1 = \begin{pmatrix} 0 & 3 & 1 \\ 3 & 1 & 0 \end{pmatrix},$$

It contains the arcs $(0 \quad 3)^T$, $(3 \quad 1)^T$, $(1 \quad 0)^T$. The cost of this contour, calculated using formula (4.28), is

$$c(Co1) = \sum_{e \in Co1} c(e) = 3.80 - 2.57 - 2.31 = -1.08.$$

The possible values $d(\eta)$ of flow's change are calculated using formula (4.29):

$$d((0 \quad 3)^T) = \beta((0 \quad 3)^T) - f((0 \quad 3)^T) = 7 - 6 = 1,$$
$$d((3 \quad 1)^T) = f((3 \quad 1)^T) = 2,$$
$$d((0 \quad 3)^T) = f((0)^T) = 4.$$

A minimal value equals 1, therefore the modified flow is the following;

$$f = (3 \quad 5 \quad 7 \quad 1 \quad 2 \quad 3 \quad 2 \quad 2 \quad 6 \quad 5 \quad 2 \quad 8).$$

The cost of new flow is smaller than the previous cost $1.08 \times 1 = 1.08$ and equals $173.94 - 1.08 = 172.85$.

Now we transfer to the second iteration and consider the following contour:

$$Co2 = \begin{pmatrix} 1 & 6 & 3 \\ 6 & 3 & 1 \end{pmatrix}.$$

The cost of this contour amounts to $2.10 - 2.57 - 3.61 = -4.08$. The possible change in the flow value is 1, giving the following flow:

$$f = (3 \quad 5 \quad 7 \quad 0 \quad 3 \quad 3 \quad 2 \quad 2 \quad 5 \quad 5 \quad 2 \quad 8)$$

The cost of the new flow equals $172.85 - 4.08 = 168.77$.

In the next iteration, the contour used is

$$Co3 = \begin{pmatrix} 2 & 5 & 7 & 4 \\ 5 & 7 & 4 & 2 \end{pmatrix}.$$

A cost of this contour comes to $3.90 + 3.35 - 4.10 - 3.55 = -0.40$. Now we have flow

$$f = (3 \quad 5 \quad 7 \quad 0 \quad 3 \quad 2 \quad 3 \quad 2 \quad 5 \quad 4 \quad 3 \quad 8),$$

having the cost of 168.37.

The contour

$$Co4 = \begin{pmatrix} 5 & 7 & 6 & 3 & 4 & 2 \\ 7 & 6 & 3 & 4 & 2 & 5 \end{pmatrix}$$

has the cost -1.157 and the flow

$$f = (3 \quad 5 \quad 7 \quad 0 \quad 3 \quad 2 \quad 3 \quad 3 \quad 4 \quad 5 \quad 3 \quad 7).$$

The cost of this flow equals $168.77 - 0.40 - 1.157 = 168.77 - 1.557 = 167.21$.

Consolidated table of results is presented in Table 4.1.

We can observe that the cost of the last presented flow is 164.82. The cost of the initial flow has decreased by 9.12.

Table 4.1. Results of sequential iterations.

Arcs, η	f	$f1$	$f2$	$f3$	$f4$	$f5$	$f6$
1	4	3	3	3	3	3	3
2	5	5	5	5	5	6	6
3	6	7	7	7	7	6	6
4	2	1	0	0	0	0	0
5	2	2	3	3	3	3	3
6	3	3	3	2	2	3	3
7	2	2	2	3	3	3	3
8	2	2	2	2	3	2	3
9	6	6	5	5	4	4	3
10	5	5	5	4	5	5	6
11	2	2	2	3	3	3	3
12	8	8	8	8	7	7	6
Cost	173.94	172.86	168.78	168.39	167.24	165.98	164.82

4.8 Conclusion

A classical problem of graph theory is considered. It is supposed that a fixed flow operates in a random environment. The latter is described as a continuous-time finite irreducible Markov chain. Each arc has the cost per unit flow transfer on that arc, which depends on the state of the random environment.

The following indices have been considered: (1) Stationary mean of the flow cost per unit time. (2) The Laplace transform of the flow cost at interval $(0, t)$, assuming a fixed initial state of the random environment. (3) Various means of the flow costs, which were calculated using this Laplace transform.

In conclusion, an optimization problem of the stationary flow is considered.

The numerical example illustrates the suggested formulas and calculating procedures. The obtained results can be augmented in various ways. For example, the correlation between flow costs for two adjacent time intervals $(0, t)$ and (t, t^*), $0 \leq t \leq t*$, can be considered. It is also possible to consider two fixed flows with different costs of arcs, operating in the same random environment.

References

[1] Hu, T.C., *Integer Programming and Network Flows.* Adisson-Wesley, California, London, Don Mills, Ontario, 1970.

[2] Cormen, T.Y., Leiserson, C.E., Rivest, R.L., Steim, C., *Introduction to Algorithms*, 2nd edn. The MIT Press, Cambridge, Massachusetts, London, 2002.

[3] Basacker, B.G., Saaty, T.L., *Finite Graphs and Networks.* McGraw-Hill, New York, 964.

[4] Brualdi, R.A., *Introductory Combinatorics*, 3rd edn. Prentice Hall, New Jersey, 1999.

[5] Christofides, N., *Graph Theory: An Algorithmic Approach.* Academic Press, New York, London, San Francisco, 1970.

[6] Ford, L.R., Fulkerson, D.R., *Flows in Networks.* Princeton University Press, Princeton, NJ, 1962.

[7] Kijima, M., *Markov Processes for Stochastic Modeling.* The University Press, Cambridge, 1997.

[8] Pacheco, A., Tang, L.Ch., Prabhu, N.U., *Markov-Modulate Processes & Semiregenerative Phenomena.* World Scientific, New Jersey, London, 2009.

[9] Råde, L., Westergren, B. *Mathematics Handbook for Science and Engineering*, 5th edn. Springer-Verlag, Berlin, Heidelberg, 2004.

Chapter 5

Wear Process Modulated by Cycling Continuous-Time Markov Chain

5.1 Introduction

Stochastic wear processes have been considered in many scientific publications on reliability and maintenance [1–6]. We consider this problem with respect to the existence of a random environment.

The process state at time t is denoted as (x, t) where $x \geq 0$ is the wear value at time $t \geq 0$. The process, which operates in a random environment, is described by a cycling continuous-time Markov chain with k states: $J(t) \in \{0, \ldots, k-1\}$, $t \geq 0$. Transitions between states occur in cycles: $0 \to 1 \to 2 \to \cdots \to k-1 \to 1 \to \cdots$, whereas transition intensities between states are given by a positive vector $\lambda = (\lambda_0 \quad \cdots \quad \lambda_{k-1})^T$. When the random environment is in the ith state, the wear velocity is $c_i \geq 0$, $i = 0, \ldots, k-1$. Thus, if the Markov chain is in the ith state during time t, the wear value increases by $c_i t$ This process stops when the accumulated wear reaches a critical level $r > 0$.

Our goal was to compute the time distribution until the critical level was reached.

The remainder of this chapter is organized as follows. An alternating Markov chain with two states is initially considered. The distribution of time until a predefined level for this alternating Markov chain is considered in Section 5.2. Section 5.3 presents a numerical example. The densities of the sojourn times in various states of a Markov chain with three states are described in Section 5.4.

The time distribution until the predefined wear level for such a chain is presented in Section 5.5. A numerical example is presented in Section 5.6. The last two sections are devoted to a case where the critical wear level r is disregarded, and the considered process continues with zero wear.

5.2 Time distribution for alternating Markov chain until a predefined level

In this section, two states of a random environment are considered: $J(\mathrm{t}) \in \{0, 1\}$, $\mathrm{t} > 0$. The transition intensities between states are given by the positive vector $\lambda = (\lambda_0 \quad \lambda_1)$. Our aim is to calculate the probability $P_i(x,t), t, x \geq 0$, that the wear $Y(t)$ at time t is less than x if the initial state of the random environment is i.

We used the results from Section 6.2. Let i and j be the initial and final states of the Markov chain $J(t)$ during the interval (0, t). The density $f_{i,j}(\tau,t)$ of the sojourn time $T_i(t)$ at the initial state i during time t jointly with the probability that the final state equals j for $0 < \tau < t$ are calculated as follows:

$$\begin{aligned} f_{i,j}(\tau,t) &= \sum_{\eta=0}^{\infty} \lambda_i \frac{1}{\eta!\eta!}(\lambda_i\, \lambda_j \tau(t-\tau))^{\eta} \exp(-t\lambda_j) \\ &\quad \times \exp(-\tau(\lambda_i - \lambda_j)), \quad j \neq i; \\ f_{i,i}(\tau,t) &= \sum_{\eta=0}^{\infty} \frac{1}{(\eta+1)!}(\tau\lambda_i)^{\eta+1} \exp(-\tau\lambda_i)\lambda_j \frac{1}{\eta!}(\lambda_j(t-\tau))^{\eta} \\ &\quad \times \exp(-(t-\tau)\lambda_j). \end{aligned} \tag{5.1}$$

The latter distribution has a singular component at point t:

$$P\{T_i(t) = t | J(0) = i\} = \exp(-\lambda_i t), \quad t \geq 0. \tag{5.2}$$

If the initial state i is fixed, the accumulated wear $Y_i(t)$ at time t is determined as follows:

$$Y_i(t) = c_i T_i(t) + c_{\mathrm{not}(i)}(t - T_i(t)), \quad t \geq 0.$$

Therefore, if the initial state of the random environment is i, the probability that the wear is less than x and that the jth state occurs at time $t \geq 0$ is calculated as follows:

$$\begin{aligned}
P_{i,j}(x,t) &= P\{Y_i(t) \leq x, J(t) = j | J(0) = i\} \\
&= P\{c_i T_i(t) + c_{\text{not}(i)}\,(t - T_i(t)) \leq x, J(t) = j | J(0) = i\} \\
&= P\{(c_i - c_{\text{not}(i)}) T_i(t) \leq x - c_{\text{not}(i)} t, J(t) = j | J(0) = i\}
\end{aligned}$$

First, we consider the case where $i \neq j$, if $\min\{c_0, c_1\}t \leq x \leq \max\{c_0, c_1\}t$, then

$$P_{i,j}(x,t) = \begin{cases} P\left\{T_i(t) \leq \dfrac{x - c_{\text{not}(i)} t}{c_i - c_{\text{not}(i)}},\ J(t) = j\right\} \\ = \displaystyle\int_0^{\frac{x - c_{\text{not}(i)} t}{c_i - c_{\text{not}(i)}}} f_{i,j}(\tau) d\tau, & \text{if } c_i > c_{\text{not}(i)}, \\ P\left\{T_i(t) \geq \dfrac{x - c_{\text{not}(i)} t}{c_i - c_{\text{not}(i)}},\ J(t) = j\right\} \\ = \displaystyle\int_{\frac{x - c_{\text{not}(i)} t}{c_i - c_{\text{not}(i)}}}^{t} f_{i,j}(\tau) d\tau, & \text{if } c_i < c_{\text{not}(i)}. \end{cases} \tag{5.3}$$

An additional singular component must be considered for the case where $i = j$. One arises because event $\{T_i(t) = t\}$ has a positive value (5.2). Let

$$\delta(x,t,i) = \begin{cases} 1 & \text{if } c_i t \leq x, \\ 0 & \text{otherwise,} \end{cases}$$

If $\min\{c_0, c_1\}t \leq x \leq \max\{c_0, c_1\}t$, we have

$$\begin{aligned}
P_{i,i}(x,t) &= P\{Y_i(t) \leq x, J(t) = i | J(0) = i\} \\
&= P\{c_i\, T_i(t) + c_{\text{not}(i)}\,(t - T_i(t)) \leq x, J(t) = j | J(0) = i\} \\
&= P\{T_i(t) = t, J(t) = i | J(0) = i\}\delta(x,t,i) + P\{c_i\, T_i(t) \\
&\quad + c_{\text{not}(i)}(t - T_i(t)) < x, J(t) = j, T_i(t) < t\}.
\end{aligned}$$

If $c_i > c_{\text{not}(i)}$, then

$$P_{i,i}(x,t) = \exp(-\lambda_i t)\delta(x,t,i) + \int_0^{\frac{x - c_{\text{not}(i)} t}{c_i - c_{\text{not}(i)}}} f_{i,i}(\tau) d\tau. \tag{5.4}$$

Conversely, if $c_i < c_{\text{not}(i))}$, then

$$P_{i,i}(x,t) = \exp(-\lambda_i t)\delta(x,t,i) + \int_{\frac{c_{\text{not}(i)}t-x}{c_{\text{not}(i)}-c_i}}^{t} f_{i,j}(\tau)d\tau. \tag{5.5}$$

Additionally, we must consider the cases where $x \leq \min\{c_0, c_1\}t$ or $x \geq \max\{c_0, c_1\}t$. Naturally if $x \leq \min\{c_0, c_1\}t$, then

$$P_{i,j}(x,t) = 0$$

The second case is considered based on the results of Chapter 1. Transition probabilities between states of the random environment $\tilde{P}_{i,j}(t) = P\{J(t) = j|J(0) = i\}$ are calculated using Eq. (1.9). The matrix of the transition intensities and the generator are

$$\boldsymbol{\lambda} = \begin{pmatrix} 0 & \lambda_{0,1} \\ \lambda_{1,0} & 0 \end{pmatrix}, \quad A = \begin{pmatrix} -\lambda_{0,1} & \lambda_{0,1} \\ \lambda_{1,0} & -\lambda_{1,0} \end{pmatrix}.$$

Given the vector of eigenvalues $\chi = (\chi_0\, \chi_1)^T$ and the matrix of eigenvectors B, the matrix of transition probabilities is calculated as follows:

$$\tilde{P}(t) = \begin{pmatrix} \tilde{P}_{0,0}(t) & \tilde{P}_{0,1}(t) \\ \tilde{P}_{1,0}(t) & \tilde{P}_{1,1}(t) \end{pmatrix} = B \begin{pmatrix} \exp(\chi_0 t) & 0 \\ 0 & \exp(\chi_1 t) \end{pmatrix} B^{-1}, \quad t \geq 0.$$

Hence, for $x > \max\{c_0, c_1\}t$, we can write

$$P_{i,j}(x,t) = \tilde{P}_{i,j}(t).$$

Finally, the probability $P_i(x,t)$ that the wear is less than x at time t if the initial state of the random environment is i for $t, x \geq 0$ can be expressed by the following equation:

$$P_i(x,t) = P_{i,0}(x,t) + P_{i,1}(x,t). \tag{5.6}$$

Note that this formula provides a distribution function of random wear $Y(t)$ if argument t is fixed. However, it provides an additional distribution function of the time until a fixed wear level x is reached.

Now, we consider the following task for optimal stopping (shut-off). Let the processes $Y_i(t)$ and $J(t)$ not be continuously observed so that only their states at initial time $t = 0$ are known, for example, the

ith state of the random environment and $Y_i(0) = 0$. If process $Y_i(t)$ does not reach the critical level r at time t and we stop it accordingly, then our reward forms t. If process $Y_i(t)$ reaches the critical level r, then there is a loss of value $l > 0$.

If the predefined stopping time is t, the expected reward $R_i(t)$ is

$$E(R_i(t,l)) = tP_i(r,t) - l(1 - P_i(r,t)), \quad t > 0. \tag{5.7}$$

The optimal stopping time t^* is the time that maximizes this expectation.

5.3 Example

Consider the following initial conditions: $c = (0.5 \;\; 1.0)^T$, $\lambda = (0.5 \;\; 1.0)^T$, $r = 4$, $n\max = 5$, which replaces the infinity ∞. Figure 5.1 shows the densities (5.1). Figure 5.2 depicts the

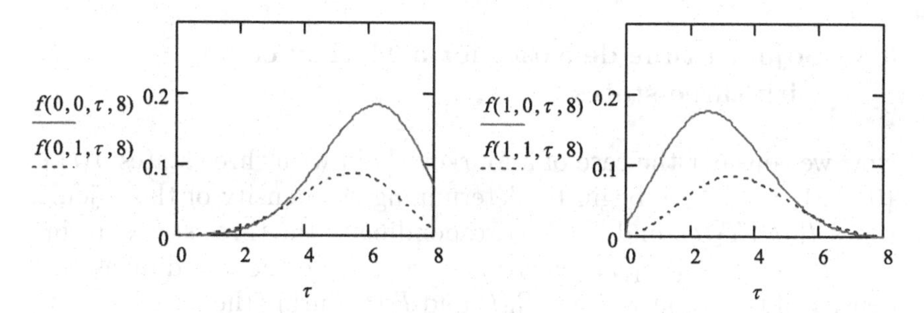

Fig. 5.1. Sojourn time densities (5.1).

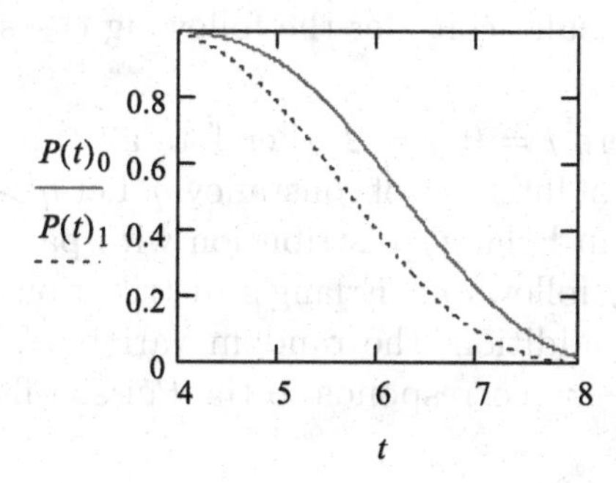

Fig. 5.2. Additional distribution function $P_i(t)$.

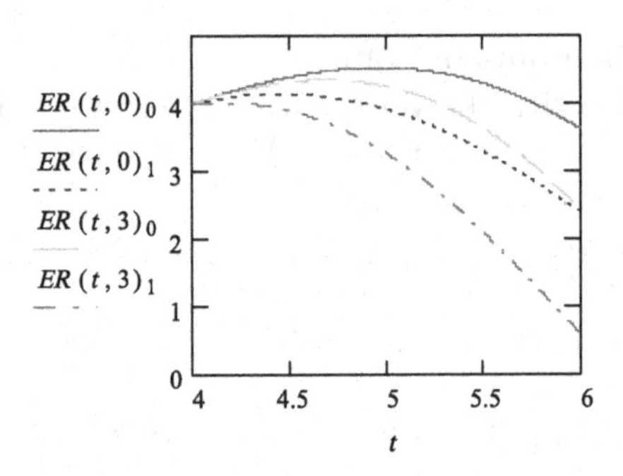

Fig. 5.3. Mean reward according to Eq. (5.7).

distribution $P_i(t) = P_{i,0}(r,t) + P_{i,1}(r,t)$ as a function of time t. Figure 5.3 plots the mean reward as a function of the predefined time t.

5.4 Sojourn time densities for a Markov chain with three states

Now we consider the case of a Markov chain with three states $J(t) \in \{0, 1, 2\}$, $t > 0$. We begin by determining the density of the sojourn times $T_0(t)$, $T_1(t)$, and $T_2(t)$ corresponding to the three states during time t. Let $f_{i,j}(\tau_0, \tau_1, t)$, where $\tau_0 + \tau_1 < t$, be the two-dimensional density of the sojourn times $T_0(t)$ and $T_1(t)$ during the interval $(0, t)$, jointly with the probability that the final state equals j if the initial state of the Markov chain is i. Then, $T_2(t) = t - T_0(t) - T_1(t)$.

We will sequentially consider the following cases:

(1) The case where $i = 0, j = 2$, therefore all three states have the same number of intervals of constancy η. Let $\eta > 0$ be fixed. Then $T_0(t)$ follows an Erlang's distribution with parameters λ_0 and η, whereas $T_1(t)$ follows an Erlang's distribution with parameters λ_1 and η. In addition, the random variable $T_2(t)$, which must be equal to $t - \tau$, corresponds to the Poisson flow with intensity

λ_2, having $\eta - 1$ arrivals during time $t - \tau_0 - \tau_1$. Therefore, for $\tau_0, \tau_1 > 0$ and $0 < \tau_0 + \tau_1 < t$:

$$
\begin{aligned}
f_{0,2}(\tau_0, \tau_1, t) &= \sum_{\eta=1}^{\infty} \prod_{\nu=0}^{1} \lambda_\nu \frac{1}{(\eta-1)!} (\tau_\nu \lambda_\nu)^{\eta-1} \exp(-\tau_\nu \lambda_\nu) \\
&\quad \times \frac{1}{(\eta-1)!} (\lambda_2 (t - \tau_0 - \tau_1))^{\eta-1} \exp(-(t - \tau_0 - \tau_1)\lambda_2) \\
&= \lambda_0 \lambda_1 \exp(-t\lambda_2) \sum_{\eta=0}^{\infty} \frac{1}{\eta!\eta!\eta!} (\tau_0 \lambda_0 \tau_1 \lambda_1 \lambda_2 (t - \tau_0 - \tau_1))^{\eta} \\
&\quad \times \exp(-(\tau_0(\lambda_0 - \lambda_2) + \tau_1(\lambda_1 - \lambda_2))), \qquad (5.8)
\end{aligned}
$$

(2) The case $i = 0, j = 0$. In this case, state 0 has the number η of its own subintervals on 1 greater than states 1 and 2. If $\eta = 1$, then the initial state 0 occurs during the entire interval $(0,\ t)$, resulting in a probability of

$$
P\{T_0(t) = t | J(0) = 0\} = \exp(-\lambda_0 t), \quad t \geq 0. \qquad (5.9)
$$

If $\eta > 1$, τ_0, τ_1 are fixed, and $\tau_0 + \tau_1 < t$, then the random variable $T_2(t)$ must be equal to $t - \tau_0 - \tau_1$. The variables $T_1(t)$ and $T_2(t)$ follow Erlang's distributions with parameters λ_1, λ_2, and $\eta - 1$. The random variable $T_0(t)$ is equal τ_0 if $\eta - 1$ arrivals of the Poisson flow with intensity λ_0 occur at time τ_0. Therefore, for $\tau_0, \tau_1 > 0$ and $0 < \tau_0 + \tau_1 < t$:

$$
\begin{aligned}
&f_{0,0}(\tau_0, \tau_1, t) \\
&= \sum_{\eta=2}^{\infty} \frac{1}{(\eta-1)!} (\tau_0 \lambda_0)^{\eta-1} \exp(-\tau_0 \lambda_0) \lambda_1 \frac{1}{(\eta-2)!} (\tau_1 \lambda_1)^{\eta-2} \exp(-\tau_1 \lambda_1) \\
&\quad \times \lambda_2 \frac{1}{(\eta-2)!} (\lambda_2 (t - \tau_1 - \tau_2))^{\eta-2} \exp(-(t - \tau_0 - \tau_1)\lambda_2) \\
&= \lambda_0 \tau_0 \lambda_1 \lambda_2 \exp(-t\lambda_2) \sum_{\eta=2}^{\infty} \frac{1}{(\eta-1)!(\eta-2)!(\eta-2)!} \\
&\quad \times^{\eta-2} \exp(-(\tau_0(\lambda_0 - \lambda_2) + \tau_1(\lambda_1 - \lambda_2))). \qquad (5.10)
\end{aligned}
$$

(3) The case $i = 0$, $j = 1$. In this case, states 0 and 1 have the number η of their own subintervals on 1 greater than state 2. If $\eta = 1$, then states 0 and 1 occur during the entire interval (0, t) considered, and the density for $T_0(t)$ is one-dimensional because $T_1(t) = t - T_0(t)$. We denote the density of $T_0(t)$ as $g(,t)$, with the argument τ jointly with probability that the state 1 occurs at time t. This density is calculated using the following equation:

$$\begin{aligned} g(\tau, t) &= \lambda_0 \exp(-\tau\lambda_0) \exp(-(t-\tau)\lambda_1) \\ &= \lambda_0 \exp(-\lambda_1 t) \exp((\lambda_1 - \lambda_0)\tau), \quad 0 < \tau < t. \end{aligned}$$

The corresponding distribution function is

$$\begin{aligned} G(\tau, t) &= P\{T_0(t) \leq \tau, J(t) = 1\} = \int_0^\tau g(\text{z}, t) dz \\ &= \int_0^\tau \lambda_0 \exp(-\lambda_1 t) \exp((\lambda_1 - \lambda_0)\text{z}) dz \\ &= \lambda_0 \exp(-\lambda_1 t) \frac{1}{\lambda_1 - \lambda_0} \{\exp((\lambda_1 - \lambda_0)\tau) - 1\}. \end{aligned}$$

Finally, we obtain

$$G(\tau, t) = \frac{\lambda_0}{\lambda_1 - \lambda_0} \exp(-\lambda_1 t)\{\exp((\lambda_1 - \lambda_0)\tau) - 1\}, \quad 0 \leq \tau \leq t. \tag{5.11}$$

The probability that the first state occurs at time t if $J(0) = 0$ is

$$P\{J(t) = 1\} = G(t, t) = \frac{\lambda_0}{\lambda_1 - \lambda_0} \{\exp(-\lambda_0 t) - \exp(-\lambda_1 \text{t})\}, \quad 0 \leq \tau \leq t. \tag{5.12}$$

If $\eta > 1$ and $\tau_0 + \tau_1 < t$, then the random variable $T_2(t)$ equals $t - \tau_0 - \tau_1$. The variables $T_0(t)$ and $T_2(t)$ follow Erlang's distributions with parameters λ_0, η, and $\lambda_2, \eta - 1$, respectively. The random variable $T_1(t)$ equals τ_1 if η arrivals of the Poisson flow with intensity λ_0 occur at time τ_1. Therefore, for τ_0, $\tau_1 > 0$ and $0 < \tau_0 + \tau_1 < t$:

$$\begin{aligned}
&f_{0,1}(\tau_0, \tau_1, t) \\
&= \sum_{\eta=2}^{\infty} \lambda_0 \frac{1}{(\eta-1)!}(\tau_0\lambda_0)^{\eta-1} \exp(-\tau_0\lambda_0) \frac{1}{\eta!}(\tau_1\lambda_1)^{\eta} \exp(-\tau_1\lambda_1) \\
&\quad \times \lambda_2 \frac{1}{(\eta-2)!}(\lambda_2(t-\tau_1-\tau_2))^{\eta-2} \exp(-(t-\tau_0-\tau_1)\lambda_2) \\
&= \tau_0\lambda_0^2(\lambda_1\tau_1)^2\lambda_2 \exp(-t\lambda_2) \sum_{\eta=2}^{\infty} \frac{1}{\eta!(\eta-1)!(\eta-2)!} \\
&\quad \times (\tau_0\lambda_0\tau_1\lambda_1\lambda_2(t-\tau_0-\tau_1))^{\eta-2} \exp(-\tau_0(\lambda_0-\lambda_2)-\tau_1(\lambda_1-\lambda_2)).
\end{aligned} \tag{5.13}$$

The above results can be verified by using the normalization condition for densities, as follows:

$$\begin{aligned}
&\int_0^t \int_0^{t-\tau_0} (f_{0,2}(\tau_0, \tau_1, t) + f_{0,1}(\tau_0, \tau_1, t) + f_{0,0}(\tau_0, \tau_1, t)) d\tau_1 d\tau_0 \\
&\quad + P\{T_0(t) = t | J(0) = 0\} + G(t,t) = 1, \quad \text{for all } t \geq 0.
\end{aligned} \tag{5.14}$$

The resulting equations suggest that the zero state of the Markov chain occurs initially: $i = 0$. For a different initial state $i \neq 0$, let α and β be the subsequent states. If $\mathrm{mod}(x, y)$ denotes the remainder after dividing x by y, then $\alpha = \mathrm{mod}(i+1, 3)$, and $\beta = \mathrm{mod}(\alpha+1, 3)$, and it is necessary to replace 0 with i, 1 with α, and 2 with β in the presented formulas.

The three cases considered are expressed as follows: case $i = 0$, $j = 2$ as $i, j = \beta$; case $i = 0, j = 0$ as $i, j = i$; and case $i = 0, j = 1$ as $i, j = \alpha$.

5.5 Time distribution until the predefined level for a Markov chain with three states

The accumulated wear $Y(t)$ at time t is determined as follows:

$$Y(t) = c_0\, T_0(t) + c_1\, T_1(t) + c_2(t - T_0(t) - T_1(t)).$$

Therefore, the distribution function of the accumulated wear during time t is

$$P(x,t) = P\{Y(t) \le x\} = P\{c_0 T_0(t) + c_1 T_1(t) + c_2(t - T_0(t) - T_1(t)) \le x\}.$$

Let $c_* = \min\{c_0, c_1, c_2\}$ and $c^* = \max\{c_0, c_1, c_2\}$. Then

$$P(x,t) = P\{Y(t) \le x\} = \begin{bmatrix} 0 & \text{if } x \le c_* t, \\ 1 & \text{if } x \ge c^* t. \end{bmatrix} \tag{5.15}$$

Regarding the initial state $i = 0$ in the interval $(0, t)$, three options are possible: (1) the initial state occurs at all times; (2) only states 0 and 1 occur; or (3) all three states occur during the considered period.

Let us consider each case, based on the following indicators:

$$\delta 0(x,t) = \begin{bmatrix} 1 & \text{if } c_0 t \le x, \\ 0 & \text{otherwise}, \end{bmatrix}$$

$$\delta 1(x,t) = \begin{bmatrix} 1 & \text{if } \max(c_0, c_1)t \le x, \\ 0 & \text{otherwise}, \end{bmatrix}$$

$$\delta 01(x,t) = \begin{bmatrix} 1 & \text{if } c_0 t \le x \le c_1 t, \\ 0 & \text{otherwise}, \end{bmatrix}$$

$$\delta 10(x,t) = \begin{bmatrix} 1 & \text{if } c_1 t \le x \le c_0 t, \\ 0 & \text{otherwise}, \end{bmatrix}$$

$$\delta(x,t,\tau_0,\tau_1) = \begin{bmatrix} 1 & \text{if } c_0\tau_0 + c_1\tau_1 + c_2(t - \tau_0 - \tau_1) \le x, \\ 0 & \text{otherwise}. \end{bmatrix}$$

If the first case takes place, then

$$P\{Y(t) \le x, J(\mathrm{t}) = 0\} = e^{-\lambda_0 t}\delta 0(t,x), \quad x,t \ge 0. \tag{5.16}$$

In the second case, states 0 and 1 occur over the entire considered interval $(0,\ t)$. The one-dimensional distribution function $G(\tau, t)$ of

$T_0(t)$ jointly with the probability that $J(t) = 1$ is calculated using Eq. (5.11). The accumulated wear $Y(t)$ for $c_0 t \le x \le \max\{c_0, c_1\}t$ is

$$\{Y(t) \le x\} = \{c_0\, T_0(t) + c_1(t - T_0(t)) \le x\}$$
$$= \{c_1 t - x \le (c_1 - c_0) T_0(t)\},$$

$$\{Y(t) \le x\} = \begin{cases} \left\{\dfrac{c_1\, t - x}{c_1 - c_0} \le T_0(t)\right\} & \text{if } c_0 < c_1, \\ \left\{\dfrac{c_1\, t - x}{c_1 - c_0} \ge T_0(t)\right\} & \text{otherwise.} \end{cases}$$

Therefore, for $c_0 < c_1$ and $c_0 t \le x \le c_1 t$:

$$P\{Y(t) \le x,\, J(\mathrm{t}) = 1\} = P\left\{\frac{c_1\, t - x}{c_1 - c_0} \le T_0(t) \le t,\, J(\mathrm{t}) = 1\right\}$$
$$= P\{T_0(t) \le t, J(\mathrm{t}) = 1\} - P\left\{T_0(t) < \frac{c_1\, t - x}{c_1 - c_0},\, J(\mathrm{t}) = 1\right\}$$
$$= G(t,t) - G\left(\frac{c_1\, t - x}{c_1 - c_0}, t\right) = \frac{\lambda_0}{\lambda_1 - \lambda_0}\{\exp(-\lambda_0 t) - \exp(-\lambda_1 t)\}$$
$$- \frac{\lambda_0}{\lambda_1 - \lambda_0} \exp(-\lambda_1 t)\left\{\exp\left((\lambda_1 - \lambda_0)\frac{c_1\, t - x}{c_1 - c_0}\right) - 1\right\}. \quad (5.17)$$

If $c_0 < c_1$ and $x \ge c_1 t$ based on (5.12) and (5.14):

$$P\{Y(t) \le x,\, J(\mathrm{t}) = 1\} = P\{J(\mathrm{t}) = 1\} = G(t,t). \quad (5.18)$$

For $c_1 < c_0$ and $c_1 t \le x \le c_0 t$:

$$P\{Y(t) \le x,\, J(\mathrm{t}) = 1\} = P\left\{\frac{c_1\, t - x}{c_1 - c_0} \ge T_0(t) \ge 0,\, J(\mathrm{t}) = 1\right\}$$
$$= P\left\{T_0(t) \le \frac{x - c_1\, t}{c_0 - c_1},\, J(\mathrm{t}) = 1\right\} = G\left(\frac{x - c_1\, t}{c_0 - c_1}, t\right). \quad (5.19)$$

If $x \ge c_0 t$, then $\{Y(t) \le x\}$ is a certain event and we have formula (5.16) again.

The third case is considered using densities (5.8), (5.10), and (5.13) only.

Finally, in addition to (5.15), the probability $P(x,t)$ can be expressed as

$$P(x,t) = e^{-\lambda_0 t}\delta(t,x) + \delta 01(t,x)\left(G(t,t) - G\left(\frac{c_1\,t - x}{c_1 - c_0}, t\right)\right)$$
$$+ \delta 10(x,t)G\left(\frac{x - c_1\,t}{c_0 - c_1}, t\right) + \delta 1(t,x)G(t,t)$$
$$+ \int_0^t \int_0^{t-\tau_0} (f_{0,2}(\tau_0,\tau_1,t) + f_{0,1}(\tau_0,\tau_1,t)$$
$$+ f_{0,0}(\tau_0,\tau_1,t))\delta(\tau_0,\tau_1,t,x)d\tau_1 d\tau_0. \tag{5.20}$$

If the predefined stopping time is t, the expectation of the obtained reward $R3(t)$ for a fixed wear r is calculated as follows:

$$E(R3(r,t,l)) = tP(r,t) - l(1 - P(r,t)), \quad t > 0. \tag{5.21}$$

Optimal stopping time is the time t that maximizes this expectation.

5.6 Example (continuation)

We consider the following initial data for a Markov chain with three states: $c = (0.5 \quad 1.0 \quad 1.5)^T$, $\lambda = (0.5 \quad 1.0 \quad 1.5)^T$, $nmax = 5$. Figure 5.4 presents the additional distribution $DAF(4,tt) = P\{Y(tt) \leq 4,\ J(tt) = 1\}$ as a function of time tt for $x = 4$, computed through Eq. (5.17). Figure 5.5 depicts the distribution function (5.17) of accumulated wear for $t = 6$, $DAF(xx,6) = P\{Y(6) \leq xx,\ J(6) = 1\}$.

Figure 5.6 contains the mean reward $E(R3(r,t,l))$ as a function of the appointed time t. The critical level r is equal to 4, and two loss values are considered: $l = 0$ and $l = 3$.

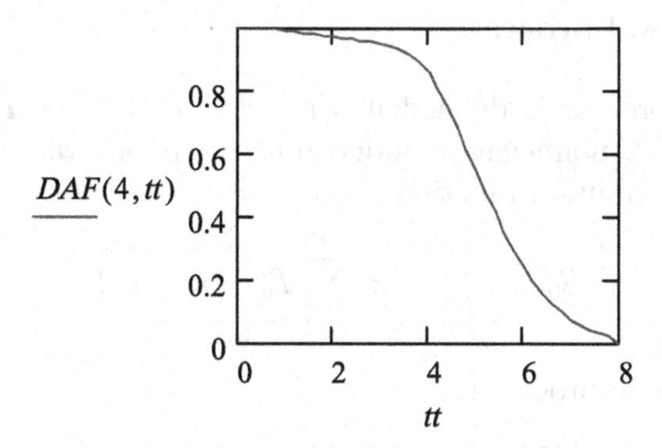

Fig. 5.4. Distribution function (5.17) for $r = 4$.

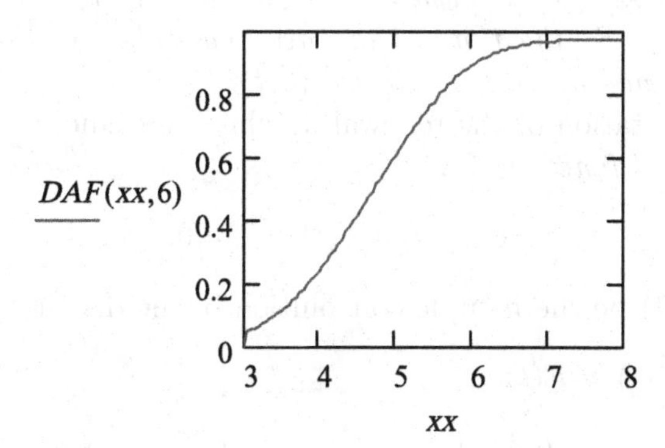

Fig. 5.5. Distribution function (5.17) for $t = 6$.

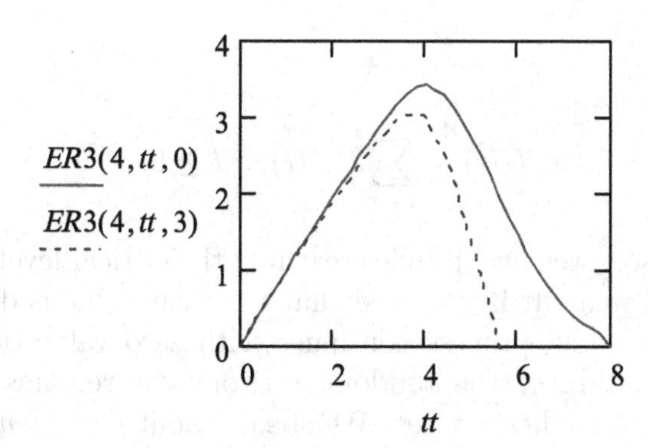

Fig. 5.6. Mean reward (5.16) for $r = 4$ as function of time *tt*.

5.7 Renewal process

A renewal process is defined in this way [7–10]. Let $T_1, T_2, \ldots$ be a sequence of nonnegative independent random variables with a common distribution function F. Let

$$S_0 = 0, \quad S_n = \sum_{\eta=1}^{n} T_\eta, \quad n = 1, 2,$$

The renewal process is

$$N(t) = \max\{n{:}S_n \leq t\}, \quad t \geq 0,$$

which indicates that *renewal* occurs at time t if $S_n = t$ for some n. We will name S_n *the time of the* nth *renewal,* and $N(t)$ *the total number of renewals on the interval* $(0, t)$.

The expectation of the renewal number as a function of time t is called *renewal function:*

$$U(t) = EN(t), \quad t \geq 0.$$

Let $F^{n*}(t)$ be the n-tuple convolution of the distribution $F(t)$:

$$F^{1*}(t) = F(t),$$

$$F^{(n+1)*}(t) = \int_0^t F^{(n)*}(t - z)dF(z), \quad t \geq 0, \ n = 1, 2, \ldots, \tag{5.22}$$

Then

$$U(t) = \sum_{n=1}^{\infty} F^{n*}(t), \quad t \geq 0. \tag{5.23}$$

In our case, a renewal implies reaching the critical level r owing to accumulated wear. It is assumed that the wear value is disregarded, and the considered process continues with zero value of the wear. Because the state of the random environment remains the same, renewals differ in these states. We speak about the *j-renewal* if the jth state occurs at the time of renewals, $j = 0, \ldots, k - 1$. Furthermore, we confine ourselves to the two-dimensional case, $k = 2$.

Let $F_{i,j}(t) = P\{T \leq t, J(t) = j|J(0) = i\}$ be the distribution function between renewals jointly with the probability that it is the *j-renewal*, if initially the ith state occurs. Distributions $P_{i,j}(x,t)$ were considered above (see formulas (5.3)–(5.5)). If $x = r$ is fixed, then it is an additional distributive function of time until this value is reached together with the probability of the final state j:

$$F_{i,j}(t) = \tilde{P}_{i,j}(t) - P_{i,j}(r,t), \quad t \geq 0, \tag{5.24}$$

where the probability $\tilde{P}_{i,j}(t) = P\{J(t) = j|J(0) = i\}$ is calculated with Eq. (1.9).

As mentioned above, the function $F_i^{n*}(t)$ is the distribution function of the time of the nth renewal if initially the ith state occurs. Convolutions $F_i^{n*}(t)$ are calculated as follows:

$$F_i^{1*}(t) = F_{i,0}(t) + F_{i,1}(t),$$

$$F_i^{(n+1)*}(t) = \sum_{\nu=0}^{k-1} \int_0^t F_\nu^{n*}(t-z)dF_{i,\nu}(z), \quad t \geq 0, \ n = 1, 2, \ldots, \tag{5.25}$$

As before, $N(t)$ denotes the total number of renewals in the interval $(0,t)$. Let the ith state take place initially, and $U_i(t)$ be the expectation of the renewal number on the interval $(0, t)$. Now, instead of (5.23), we have

$$U_i(t) = \sum_{n=1}^{\infty} F_i^{n*}(t), \quad t \geq 0. \tag{5.26}$$

Let us make some remarks concerning this numerical implementation. Difficulties arise with repeated calculations of convolutions $F_{i,j}^{(n)*}(t)$. To overcome that, we use a set of lattice points and substitute integrals with sums. The continuous time is replaced with discrete time steps Δ, which is decreased until changes are significant.

The iterative procedure in (5.25) uses four tables to maintain the intermediate data. The first table contains values of the increasing $dF_{i,\nu}(z)$ at one step. The second table contains the values $F_i^{(n)*}(t)$ of the previous multiplicity n. Initially, the second table contains the values of $F_i^{1*}(t)$. The third table lists the values of the multiplicity $n+1$. After a complication of this table's calculation, this table becomes the second table, and the iterative procedure continues. The final table U, which is useful for calculating the renewal function (5.26), contains two rows corresponding to the initial states, with numbers $i = 0$ and $i = 1$.

5.8 Example (continuation)

We continue to consider the first example. The original data is the follows: $c = (0.5 \quad 1.0)^T$, $\lambda = (0.5 \quad 1.0)^T$, and the critical level $r = 2$. First, we consider the transition probabilities $\tilde{P}_{i,j}(t) = P\{J(t) = j|J(0) = i\}$ between the states of the random environment. In our case,

$$\boldsymbol{\lambda} = \begin{pmatrix} 0 & 0.5 \\ 1.0 & 0 \end{pmatrix}, \quad A = \begin{pmatrix} -0.5 & 0.5 \\ 1.0 & -1.0 \end{pmatrix}, \quad \chi = \begin{pmatrix} 0 \\ -1.5 \end{pmatrix},$$

$$B = \begin{pmatrix} 0.707 & -0.447 \\ 0.707 & 0.894 \end{pmatrix}, \quad B^{-1} = \begin{pmatrix} 0.943 & 0.471 \\ -0.745 & 0.745 \end{pmatrix}.$$

The matrix of transition probabilities is as follows:

$$\begin{aligned} \tilde{P}(t) &= \begin{pmatrix} \tilde{P}_{0,0}(t) & \tilde{P}_{0,1}(t) \\ \tilde{P}_{1,0}(t) & \tilde{P}_{1,1}(t) \end{pmatrix} \\ &= \begin{pmatrix} 0.707 & -0.447 \\ 0.707 & 0.894 \end{pmatrix} \begin{pmatrix} 1 & 0 \\ 0 & \exp(-1.5t) \end{pmatrix} \begin{pmatrix} 0.943 & 0.471 \\ -0.745 & 0.745 \end{pmatrix} \\ &= \begin{pmatrix} 0.667 + 0.333 * \exp(-1.5t) & 0.333 - 0.333 * \exp(-1.5t) \\ 0.667 - 0.667 * \exp(-1.5t) & 0.333 + 0.667 * \exp(-1.5t) \end{pmatrix}, \end{aligned}$$

$$t \geq 0. \tag{5.27}$$

This matrix is presented in Table 5.1, where the time step Δ equals 0.1 and the number $\eta = 2i + j$ of the row defines a transition from state i to state j. The values of the matrix for $\theta > 29$ are close to the stationary distribution $\tilde{P}_{i,0}(\infty) = 0.666$, $\tilde{P}_{i,1}(\infty) = 0.334$, $i = 0, 1$.

The functions $\Pr(t, j)_i = P_{i,j}(r, t)$ and $\text{PAll}(t)_i = \Pr(t, 0)_i + \Pr(t, 1)_i$ for $r = 2$ as a function of time t are presented in Figure 5.7.

The same data is presented in the Table 5.2, namely, the probabilities that the wear is less than $r = 2$ and the jth state occurs at time $t = \Delta\,\theta = 0.1\theta$. Row zero corresponds to the transition of the random environment from state $J = 0$ to state $J = 0$. The last column of the table shows that $P_{0,0}(2, 4) = 0.135$. If $t > 4$, then $P_{0,0}(2, t) = 0$. Therefore, 0.135 is the value of the jump in the probabilistic distribution at time $t = 4$. This result can be verified as follows. The critical

Table 5.1. Matrix $\tilde{P}(t)$ for $t = 0.1\theta$.

θ	0	1	2	3	4	5	6	7	8	9
$\eta = 0$	1	0.954	0.914	0.879	0.850	0.824	0.802	0.783	0.767	0.753
$\eta = 1$	0	0.046	0.914	0.121	0.150	0.176	0.198	0.217	0.233	0.247
$\eta = 2$	1	0.093	0.086	0.242	0.301	0.352	0.396	0.433	0.466	0.494
$\eta = 3$	0	0.907	0.173	0.758	0.699	0.648	0.604	0.567	0.534	0.506
θ	10	11	12	13	14	15	16	17	18	19
$\eta = 0$	0.741	0.731	0.722	0.714	0.7 07	0.702	0.697	0.693	0.689	0.686
$\eta = 1$	0.259	0.269	0.278	0.286	0.293	0.298	0.303	0.307	0.311	0.314
$\eta = 2$	0.518	0.539	0.556	0.572	0.585	0.596	0.606	0.615	0.622	0.628
$\eta = 3$	0.482	0.461	0.444	0.428	0.415	0.404	0.394	0.385	0.378	0.372
θ	20	21	22	23	24	25	26	27	28	29
$\eta = 0$	0.683	0.681	0.679	0.677	0.676	0.675	0.673	0.672	0.672	0.671
$\eta = 1$	0.317	0.319	0.321	0.323	0.324	0.325	0.327	0.328	0.328	0.329
$\eta = 2$	0.633	0.638	0.642	0.646	0.648	0.651	0.653	0.655	0.657	0.658
$\eta = 3$	0.367	0.352	0.358	0.354	0.352	0.349	0.347	0.345	0.343	0.342

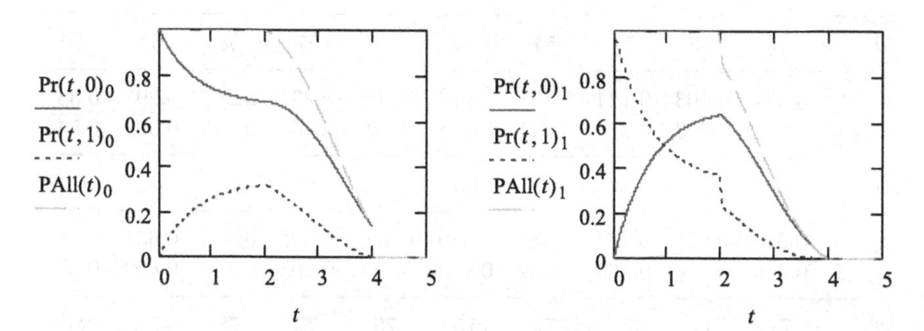

Fig. 5.7. The graphics of the function $P_{i,j}(2, t)$.

level $r = 2$ is first reached at instant $t = r/c_0 = 2/0.5 = 4$, indicating that the initial zero state continuously occurs until $t = 4$. The corresponding probability is

$$\exp(-\lambda_{0,0}4) = \exp(-0.5 \times 4) = 0.135.$$

An analogous analysis can be performed for a case where the initial first state occurs continuously until instant $t = r/c_1 = 2$. The probability that the critical level $r = 2$ is first reached at

Table 5.2. Probabilities $P_{i,j}(2,t)$ for $t = 0.1\theta$.

θ	20	21	22	23	24	25	26	27	28	29
$\eta = 0$	0.683	0.680	0.673	0.664	0.652	0.636	0.617	0.595	0.570	0.541
$\eta = 1$	0.317	0.305	0.291	0.276	0.260	0.242	0.224	0.204	0.184	0.164
$\eta = 2$	0.633	0.610	0.583	0.552	0.520	0.484	0.447	0.408	0.369	0.329
$\eta = 3$	0.367	0.212	0.193	0.174	0.155	0.136	0.118	0.101	0.085	0.070
30	31	32	33	34	35	36	37	38	39	40
0.510	0.476	0.440	0.403	0.364	0.324	0.285	0.245	0.207	0.170	0.135
0.144	0.125	0.106	0.088	0.071	0.055	0.041	0.028	0.017	0.008	0.000
0.289	0.212	0.583	0.176	0.142	0.110	0.081	0.056	0.034	0.015	0.000
0.057	0.034	0.193	0.025	0.018	0.012	0.007	0.004	0.002	0.000	0.000

Table 5.3. Probabilities $F_{i,j}^{(2)*}(t)$ for $t = 0.1\theta$.

θ	41	42	43	44	45	46	47	48	49	50
$i = 0$	0.000	0.002	0.005	0.010	0.015	0.022	0.031	0.041	0.053	0.068
$i = 1$	0.000	0.028	0.039	0.053	0.068	0.086	0.107	0.129	0.154	0.182
θ	51	52	53	54	55	56	57	58	59	60
$i = 0$	0.084	0.103	0.124	0.147	0.173	0.201	0.231	0.263	0.297	0.333
$i = 1$	0.211	0.243	0.276	0.311	0.348	0.386	0.425	0.465	0.506	0.546
θ	61	62	63	64	65	66	67	68	69	70
$i = 0$	0.370	0.407	0.444	0.482	0.519	0.556	0.591	0.625	0.657	0.688
$i = 1$	0.586	0.626	0.665	0.702	0.738	0.771	0.803	0.832	0.859	0.883
θ	71	72	73	74	75	76	77	78	79	80
$i = 0$	0.716	0.742	0.765	0.786	0.805	0.820	0.834	0.845	0.853	1.000
$i = 1$	0.905	0.924	0.940	0.955	0.967	0.976	0.984	0.990	0.995	1.000

this instant is 0.155. This follows from the last row of Table 5.2, where $P_{1,1}(2,2) = 0.367$, $P_{1,1}(2,2.1) = 0.212$. The difference is 0.155.

Table 5.3 contains the convolutions $F_i^{(2)*}(t)$ defined by (5.25). The values $dF_{i,j}(z)$ were calculated as the differences between the

Table 5.4. Renewal functions $U_i(t)$ for $t = 0.1\theta$.

θ	20	22	24	26	28	30	32	34	36	38
$i = 0$	0.000	0.035	0.089	0.159	0.246	0.346	0.454	0.565	0.675	0.776
$i = 1$	0.000	0.224	0.325	0.435	0.546	0.654	0.754	0.841	0.911	0.965
θ	40	42	44	46	48	50	52	54	56	58
$i = 0$	0.865	1.002	1.010	1.022	1.041	1.068	1.103	1.147	1.201	1.263
$i = 1$	1.000	1.028	1.053	1.086	1.129	1.182	1.243	1.311	1.386	1.465
θ	60	62	64	66	68	70	72	74	76	78
$i = 0$	1.333	1.407	1.483	1.588	1.631	1.699	1.762	1.818	1.869	1.914
$i = 1$	1.546	1.626	1.709	1.785	1.856	1.922	1.982	2.038	2.090	2.142
θ	80	82	84	86	88	90	92	94	96	98
$i = 0$	1.956	2.129	2.167	2.211	2.260	2.313	2.368	2.425	2.482	2.537
$i = 1$	2.193	2.245	2300	2.360	2.422	2.486	2.549	2.610	2.668	2.720
θ	100	102	104	106	108	110	112	114	116	118
$i = 0$	2.588	2.634	2.677	2.714	2.746	2.774	2.798	2.818	2.834	2.848
$i = 1$	2.766	2.808	2.847	2.882	2.911	2.935	2.455	2.971	2.983	2.992

corresponding values in Table 5.2. Table 5.4 contains the values of the renewal function (5.26).

The above numerical results indicate that an elaborate approach to the analysis of the wear processes facilitates taking correct decisions regarding their functions.

References

[1] Beihelt, F., Franken P., *Zuferlassigkeit und instandhaltung. Mathematische methoden.* VEB Verlag Technik, Berlin, 1983.

[2] Blau, P.J., *Tribosystem Analysis — A Practical Approach to the Diagnosis of Wear Problems.* CRC Press, Boca Raton, 2016.

[3] Bovaird, R.I., Characteristics of optimal maintenance politics, *Management Science.* 1961;7:238–253.

[4] Gertsbakh, I., *Reliability Theory. With Applications to Preventive Maintenance.* Springer, 2000.

[5] Gertsbakh, I., Kordonsky, Kh., *Models of Failures.* Berlin, Heidelberg, Springer, New York, 1968.

[6] Mercer, A., On wear-depending renewal processes. *Journal Royal Statistics Society B*, 1961;23:368–376.

[7] Smith, W., Renewal theory and its ramifications. *Journal of the Royal Statistical Society, Series B*, 1958;20(2):243–302.

[8] Cox, D.R., *Renewal Theory.* Methuen and Co. Ltd. – John London, Wiley & Sons Inc. New York, 1995.

[9] Feller, W., *An introduction to Probability Theory and its Applications.* volume II, 2nd edn., Wiley, New York, 1971.

[10] Ross, Sh.M., *Applied Probability Model with Optimization Applications.* Dover Publication, New York, 1991.

Chapter 6

Alternating Poisson Processes and Their Estimation

6.1 Introduction

In this chapter, we expand on the Poisson process considering Example 1.4 and Section 2.2. We now interpret the process as the flow of random arriving events. The value of the process at instant t indicates the *number of arrived events within the interval* $(0, t)$.

Poisson flows are used widely in probabilistic models for various applications, including but not limited to reliability analysis, queuing theory, storage systems, risk assessment, and transportation networks [1–3]. A generalization of the Poisson flow is a *Markov-modulated Poisson process* (MMPP), in which the intensity of the Poisson flow depends on the state of *an external random environment.* Usually, this environment is described as *a continuous-time Markov chain* (MC) [4–6]. The MMPP has wide utility in problems such as the transmission of signal and information, reliability of complex devices and software, and maintenance policies [7,8].

In this chapter, we examine the simple case of an external random environment. The latter is assumed to be an alternating continuous-time MC with two states. This enables an efficient calculation procedure that does not require matrix exponents.

A Poisson flow of arrivals is considered when the flow intensity is not constant. The flow operates in a random environment that is described as an alternating MC $J(t)$. The sojourn times in the two alternating states are independent random variables having exponential distributions with parameters λ_1 and λ_2, respectively.

The intensity of the flow is α_i, $i = 1, 2$, if the ith state of the random environment takes place.

The following indices of the flow are considered: the distribution, expectation, and variance of the number of arrivals in a given interval, and the correlation of the number of arrivals for two adjacent intervals.

The remainder of this chapter is organized as follows. The distribution of the sojourn time in the fixed state of the MC is derived in Section 6.2. Section 6.3 presents the analysis of one flow in a single time interval. The case of two adjacent intervals is discussed in Section 6.4. Two or more flows are considered in Section 6.5. Section 6.6 introduces a numerical example. Some estimation procedures are described in Sections 6.7 and 6.8. Lastly, Section 6.9 culminates with the final remarks. The necessary proof is provided in Appendix.

6.2 Distribution of sojourn time in the fixed state of MC $J(t)$

Let i and j be the initial and final states of the MC in interval $(0, t)$. We wish to calculate the density $f_{i,j}(\tau, t)$ of the sojourn time $T_i(t)$ at point τ for the initial state i during time t jointly with the probability that the final state equals j. The following possible cases must be considered:

(1) $i \neq j$. In this case, both states have the same number η of constancy intervals. If $\eta > 0$ is fixed, then $T_i(t)$ follows an Erlang distribution with parameters λ_i and η. If $T_i(t) = \tau < t$, then the random variable $T_j(t)$ must be equal to $t - \tau$. It takes place if the Poisson flow with intensity λ_j has $\eta - 1$ arrivals during time $t - \tau$. Therefore, for $0 < \tau < t$,

$$\begin{aligned} f_{i,j}(\tau, t) &= \sum_{\eta=1}^{\infty} \lambda_i \frac{1}{(\eta-1)!}(\tau\lambda_i)^{\eta-1}\exp(-\tau\lambda_i)\frac{1}{(\eta-1)!} \\ &\quad \times (\lambda_j(t-\tau))^{\eta-1}\exp(-(t-\tau)\lambda_j) \\ &= \sum_{\eta=0}^{\infty} \lambda_i \frac{1}{\eta!\eta!}(\lambda_i\lambda_j\tau(t-\tau))^{\eta}\exp(-t\lambda_j) \\ &\quad \times \exp(-\tau(\lambda_i - \lambda_j)), \quad j \neq i. \end{aligned} \tag{6.1}$$

(2) $i = j$. In this case, state i has 1 more subinterval than state $not(i)$. If $\eta = 1$, then initial state i occurs during the entire considered interval $(0, t)$, with a corresponding probability of $\exp(-\lambda_i t)$. If $\tau < t$ and $\eta > 1$ are fixed, then $T_{not(i)}(t)$ follows an Erlang distribution with parameters $\lambda_{not(i)}$ and η. The random variable $T_i(t)$ must be equal to $t - T_{not(i)}(t)$. It occurs if the Poisson flow with intensity λ_i has $\eta + 1$ arrivals during time $t - T_{not(i)}(t)$. Therefore,

$$f_{i,i}(\tau, t) = \sum_{\eta=2}^{\infty} \frac{1}{(\eta-1)!}(\tau\lambda_i)^{\eta-1} \exp(-\tau\lambda_i)\lambda_{\text{not}(i)}$$

$$\times \frac{1}{(\eta-2)!}(\lambda_{\text{not}(i)}(t-\tau))^{\eta-2} \exp(-(t-\tau)\lambda_{\text{not}(i)})$$

$$= \sum_{\eta=0}^{\infty} \frac{1}{(\eta+1)!}(\tau\lambda_i)^{\eta+1} \mathrm{e}^{-\tau\lambda_i} \lambda_{\text{not}(i)} \frac{1}{\eta!}(\lambda_{\text{not}(i)}(t-\tau))^{\eta}$$

$$\times \mathrm{e}^{-(t-\tau)\lambda_{\text{not}(i)}}. \tag{6.2}$$

The last distribution has a singular component at point t:

$$P\{T_i(t) = t | J(0) = i\} = \exp(-\lambda_i t), \ t \geq 0.$$

We can now calculate the expectation of the sojourn time in the initial state i:

$$E(T_i(t) | J(0) = i) = \int_0^t \tau(f_{i,\text{not}(i)}(\tau, t) + f_{i,i}(\tau, t))d\tau + t\exp(-\lambda_i t). \tag{6.3}$$

There is an alternative expression for computing the expectation. It is based on the transition probability $\Pr_{i,j}(t)$ that the random environment has state j at time t, if the ith state occurs first. Let $\Pr(t) = (\Pr_{i,j}(t))$ correspond to a 2×2 matrix. In this case, we have the following *generator* (see Section 1.2):

$$A = \begin{pmatrix} -\lambda_1 & \lambda_1 \\ \lambda_2 & -\lambda_2 \end{pmatrix}.$$

Let $\chi = (\chi_1 \ \chi_2)^{\mathrm{T}}$ be the vector of eigenvalues and $B = (\beta_1 \ \beta_2)$ the matrix of eigenvectors of A. Then,

$$\Pr(t) = B\text{diag}((\exp(t\chi_1) \ \exp(t\chi_2)))B^{-1}.$$

The alternative approach is as follows:

$$E(T_i(t)|J(0) = i) = \int_0^t \Pr_{i,i}(t)d\tau.$$

The probability $q_i(t)$ that the initial state i occurs at the end of the interval $(0, t)$ is given by the diagonal element of $\Pr(t)$:

$$q_i(t) = \Pr_{i,i}(t). \tag{6.4}$$

6.3 One Poisson flow

An arriving Poisson flow has intensity α_i if the external random environment has state $i = 1, 2$. Let $X(0, t)$ denote the number of arrivals in the interval $(0, t)$. Equations (6.1)–(6.3) allow for the calculation of the probability $P_{i,j}(n, t)$ of n arrivals during interval $(0, t)$ and the state j at instant t if the state i occurs initially.

If sojourn time $T_i(t)$ equals $\tau < t$, the number of arrivals of two Poisson flows are independent and their sum follows a Poisson distribution. Therefore,

$$\begin{aligned}
P_{i,j}(n,t) = &\int_0^t f_{i,j}(\tau,t)\frac{1}{n!}(\alpha_i\tau + \alpha_j(t-\tau))^n \\
&\times \exp(-(\alpha_i\tau + \alpha_j(t-\tau)))d\tau, \quad i \neq j, \quad n = 0, 1, \ldots, \qquad (6.5)\\
P_{i,i}(n,t) = &\int_0^t f_{i,i}(\tau,t)\frac{1}{n!}(\alpha_i\tau + \alpha_{not(i)}(t-\tau))^n \\
&\times \exp(-(\alpha_i\tau + \alpha_{not(i)}(t-\tau)))d\tau \\
&+ \exp(-t\lambda_i)\frac{1}{n!}(\alpha_i t)^n \exp(-\alpha_i t), \quad n = 0, 1, \ldots. \qquad (6.6)
\end{aligned}$$

Finally, the probability $\Pr_i(n, t)$ of n arrivals within the interval $(0, t)$ is calculated as

$$\Pr_i(n,t) = P_{i,1}(n,t) + P_{i,2}(n,t), \quad t \geq 0, \ n = 0, 1, \ldots. \tag{6.7}$$

Furthermore, various numerical indices are calculated. For example, the expectation jointly with the probability that the last state equals $j \neq i$ is given by

$$
\begin{aligned}
E_{i,j}(X(0,t)) &= \int_0^t f_{i,j}(\tau,t)(\alpha_i\tau + \alpha_j(t-\tau))d\tau \\
&= \sum_{\eta=0}^{\infty}\int_0^t \lambda_i \frac{1}{\eta!\eta!}(\lambda_i\lambda_j\tau(t-\tau))^{\eta} \\
&\quad \times \exp(-t\lambda_j)\exp(-\tau(\lambda_i-\lambda_j))(\alpha_i\tau+\alpha_j(t-\tau))d\tau \\
&= \alpha_i\lambda_i\exp(-t\lambda_j)\sum_{\eta=0}^{\infty}\int_0^t \frac{1}{\eta!\eta!}(\lambda_i\lambda_j(t-\tau))^{\eta}\tau^{\eta+1} \\
&\quad \times \exp(-\tau(\lambda_i-\lambda_j))d\tau \\
&\quad + \alpha_j(\lambda_j)^{-1}\exp(-t\lambda_j)\sum_{\eta=0}^{\infty}\int_0^t \frac{1}{\eta!\eta!}(\lambda_i\lambda_j(t-\tau))^{\eta+1}\tau^{\eta} \\
&\quad \times \exp(-\tau(\lambda_i-\lambda_j))d\tau.
\end{aligned}
$$

For $i = j$, the expectation is expressed as

$$
\begin{aligned}
E_{i,i}(X(0,t)) &= \int_0^t f_{i,i}(\tau,t)(\alpha_i\tau + \alpha_{\mathrm{not}(i)}(t-\tau))d\tau + \alpha_i t\exp(-t\lambda_i) \\
&= \int_0^t \sum_{\eta=0}^{\infty}\frac{1}{(\eta+1)!}(\tau\lambda_i)^{\eta+1}\ \exp(-\tau\lambda_i)\lambda_{\mathrm{not}(i)}\frac{1}{\eta!} \\
&\quad \times(\lambda_{\mathrm{not}(i)}(t-\tau))^{\eta}\exp(-(t-\tau)\lambda_{\mathrm{not}(i)}) \\
&\quad \times(\alpha_i\tau+\alpha_{\mathrm{not}(i)}(t-\tau))d\tau + \alpha_i t\exp(-t\lambda_i) \\
&= \alpha_i\lambda_i\exp(-t\lambda_{\mathrm{not}(i)})\sum_{\eta=0}^{\infty}\int_0^t \frac{1}{(\eta+1)!\eta!} \\
&\quad \times(\lambda_i\lambda_{\mathrm{not}(i)}(t-\tau))^{\eta}\tau^{\eta+2}\exp(-\tau(\lambda_i-\lambda_{\mathrm{not}(i)}))d\tau \\
&\quad + \alpha_{\mathrm{not}(i)}\exp(-t\lambda_{\mathrm{not}(i)})\sum_{\eta=0}^{\infty}\int_0^t \frac{1}{(\eta+1)!\eta!} \\
&\quad \times(\lambda_i\lambda_{\mathrm{not}(i)}\tau(t-\tau))^{\eta+1} \\
&\quad \times\exp(-\tau(\lambda_i-\lambda_{\mathrm{not}(i)}))d\tau + \alpha_i t\exp(-t\lambda_i).
\end{aligned}
$$

Finally, for $i \neq j$,

$$
\begin{aligned}
E_{i,j}(X(0,t)) = {} & \alpha_i \lambda_j \exp(-t\lambda_j) \sum_{\eta=0}^{\infty} \int_0^t \frac{1}{\eta!\eta!} (\lambda_i \lambda_j (t-\tau))^{\eta} \tau^{\eta+1} \\
& \times \exp(-\tau(\lambda_i - \lambda_j)) d\tau \\
& + \alpha_j (\lambda_j)^{-1} \exp(-t\lambda_j) \sum_{\eta=0}^{\infty} \int_0^t \frac{1}{\eta!\eta!} (\lambda_i \lambda_j (t-\tau))^{\eta+1} \tau^{\eta} \\
& \times \exp(-\tau(\lambda_i - \lambda_j)) d\tau; \qquad (6.8) \\
E_{i,i}(X(0,t)) = {} & \alpha_i t \exp(-t\lambda_i) + \alpha_i \lambda_i \exp(-t\lambda_{\text{not}(i)}) \\
& \times \sum_{\eta=0}^{\infty} \int_0^t \frac{1}{(\eta+1)!\eta!} (\lambda_i \lambda_{\text{not}(i)} (t-\tau))^{\eta} \tau^{\eta+2} \\
& \times \exp(-\tau(\lambda_i - \lambda_{\text{not}(i)})) d\tau + \alpha_{\text{not}(i)} \exp(-t\lambda_{\text{not}(i)}) \\
& \times \sum_{\eta=0}^{\infty} \int_0^t \frac{1}{(\eta+1)!\eta!} (\lambda_i \lambda_{\text{not}(i)} \tau (t-\tau))^{\eta+1} \\
& \times \exp(-\tau(\lambda_i - \lambda_{\text{not}(i)})) d\tau. \qquad (6.9)
\end{aligned}
$$

The unconditional expectation is as follows:

$$
E_i(X(0,t)) = E_{i,1}(X(0,t)) + E_{i,2}(X(0,t)). \qquad (6.10)
$$

The second moment is calculated analogously. For $i \neq j$,

$$
\begin{aligned}
E_{i,j}(X(0,t)^2) &= \sum_{n=1}^{\infty} n^2 P_{i,j}(n,t) \\
&= \lambda_i \exp(-t\lambda_j) \sum_{\eta=0}^{\infty} \frac{1}{\eta!\eta!} (\lambda_i \lambda_j)^{\eta} \\
&\quad \times \int_0^t (\tau(t-\tau))^{\eta} [(\alpha_i \tau + \alpha_j (t-\tau))^2 + \alpha_i \tau \\
&\qquad + \alpha_j (t-\tau)] \exp(-\tau(\lambda_i - \lambda_j)) d\tau; \qquad (6.11)
\end{aligned}
$$

$$E_{i,i}(X(0,t)^2) = \sum_{n=1}^{\infty} n^2 P_{i,i}(n,t)$$

$$= \exp(-t\lambda_{\text{not}(i)}) \sum_{\eta=1}^{\infty} \frac{1}{(\eta+1)!\eta!} (\lambda_i \lambda_{\text{not}(i)})^{\eta+1}$$

$$\times \int_0^t (t-\tau)^{\eta} \tau^{\eta+1} [(\alpha_i \tau + \alpha_{\text{not}(i)}(t-\tau))^2$$

$$+ (\alpha_i \tau + \alpha_{\text{not}(i)}(t-\tau))] \times \exp(-\tau(\lambda_i - \lambda_{\text{not}(i)}))d\tau$$

$$+ [(\alpha_i t)^2 + \alpha_i t] \exp(-t\lambda_i). \quad (6.12)$$

The unconditional second moment is as follows:

$$E_i(X(0,t)^2) = E_{i,1}(X(0,t)^2) + E_{i,2}(X(0,t)^2). \quad (6.13)$$

The variance and standard deviation are calculated as usual:

$$\text{Var}_i(X(0,t)) = E_i(X(0,t)^2) - (E_i(X(0,t)))^2, \quad (6.14)$$

$$\sigma_i(X(0,t)) = \sqrt{\text{Var}_i(X(0,t))}. \quad (6.15)$$

6.4 One Poisson flow on two adjacent internals

We now consider two adjacent intervals: $(0,t)$ and (t,t^*), $0 < t < t^*$. The joint distribution $P_i(n_1,n_2,t,t^*) = P\{X(0,t) = n_1, X(t,t^*) = n_2 | J(0) = i\}$ is as follows:

$$P_i(n_1,n_2,t,t^*) = \sum_{j=1}^{2} P_{i,j}(n_1,t)\text{Pr}_j(n_2,t^*-t), \quad n_1,n_2 = 0,1,\ldots. \quad (6.16)$$

The second mixed moment is calculated as

$$E_i(X(0,t)X(t,t^*)) = \sum_{n_1=1}^{\infty}\sum_{n_2=1}^{\infty} n_1 n_2 P_{i,i}(n_1,t)\text{Pr}_i(n_2,t^*-t)$$

$$+ \sum_{n_1=1}^{\infty}\sum_{n_2=1}^{\infty} n_1 n_2 P_{i,\text{not}(i)}(n_1,t)\text{Pr}_{\text{not}(i)}(n_2,t^*-t)$$

$$= E_{i,i}(X(0,t))E_i(X(t,t^*)) + E_{i,\text{not}(i)}(X(0,t))$$

$$\times E_{\text{not}(i)}(X(t,t^*)), \quad 0 < t < t^*. \quad (6.17)$$

Now we can calculate the covariance using formula (6.4):

$$\begin{aligned}\mathrm{Cov}_i(X(0,t)X(t,t^*)) &= E_i(X(0,t)X(t,t^*)) - E_i(X(0,t))\\ &\quad\times (q_i(t)E_i(X(t,t^*)) + (1-q_i(t))\\ &\quad\times E_{\mathrm{not}(i)}(X(t,t^*))), \quad 0<t<t^*,\ i=1,2.\end{aligned} \tag{6.18}$$

Next, to calculate the correlation coefficients, we need the variance $\mathrm{Var}(X(t,t^*)|J(0) = i)$, which can be computed using the following expression (see Appendix):

$$\begin{aligned}\mathrm{Var}(X(t,t^*|J(0)=i) &= q_i(t)\mathrm{Var}_i(X(t,t^*))\\ &\quad + (1-q_i(t))\mathrm{Var}_{\mathrm{not}(i)}(X(t,t^*)\\ &\quad + q_i(t)(1-q_i(t))(E_i(X(t,t^*))\\ &\quad - E_{\mathrm{not}(i)}(X(t,t^*)))^2.\end{aligned} \tag{6.19}$$

Therefore,

$$\begin{aligned}\rho_i(t) &= \mathrm{Cov}_i(X(0,t)X(t,t^*))\\ &\quad\times\sigma_i(X(0,t))^{-1}(\mathrm{Var}(X(t,t^*)|J(0)=i))^{-\frac{1}{2}}.\end{aligned} \tag{6.20}$$

6.5 Many Poisson flows

Let us now assume that k Poisson flows occur. If the state i of the external environment is fixed, the arrivals of different flows form independent stochastic processes. The intensity of the νth flow equals $\alpha_{i,\nu}$. Let $\alpha_i = (\alpha_{i,1},\ldots,\alpha_{i,k})$ be the vector of intensities, $\vec{X}(t,t^*) = (X_1(t,t^*),\ldots,X_k(t,t^*))$ the vector of the number of arrivals of the flows, R^k the k-dimensional lattice of the nonnegative integers, and $\vec{n} = (n_1,\ldots,n_k) \in R^k$ the vector of variables.

The marginal distribution for the νth flow is calculated by replacing α_i by $\alpha_{i,\nu}$ in Eqs. (6.5)–(6.15). The joint distribution is given by the generalization of formulas (6.5) and (6.6) for $i \neq j$, $\vec{n} \in R^k$:

$$P_{i,j}(\vec{n},t) = P\{\vec{X}(0,t) = \vec{n},\ J(t) = j|J(0) = i\}$$
$$= \int_0^t f_{i,j}(\tau,t) \prod_{\nu=1}^{k} \frac{1}{n_\nu!}(\alpha_{i,\nu}\tau + \alpha_{j,\nu}(t-\tau))^{n_\nu}$$
$$\times \exp(-(\alpha_{i,\nu}\tau + \alpha_{j,\nu}(t-\tau)))d\tau; \tag{6.21}$$

$$P_{i,i}(\vec{n},t) = \int_0^t f_{i,i}(\tau,t) \prod_{\nu=1}^{k} \frac{1}{n_\nu!}(\alpha_{i,\nu}\tau + \alpha_{\text{not}(i),\nu}(t-\tau))^{n_\nu}$$
$$\times \exp(-(\alpha_{i,\nu}\tau + \alpha_{\text{not}(i),\nu}(t-\tau)))d\tau$$
$$+ \exp(-t\lambda_i) \prod_{\nu=1}^{k} \frac{1}{n_\nu!}(\alpha_{i,\nu}t)^{n_\nu} \exp(-\alpha_{i,\nu}t), \quad \vec{n} \in R^k. \tag{6.22}$$

The second mixed moment of the number of arrivals of the first and second flows is as follows:

$$E_i(X_1(0,t)X_2(0,t)) = E(X_1(0,t)X_2(0,t)|J(0) = i)$$
$$= \exp(-\lambda_i t)t^2\alpha_{i,1}\alpha_{i,2}$$
$$+ \sum_{j=1}^{2} \int_0^t f_{i,j}(\tau,t)(\alpha_{i,1}\tau + \alpha_{j,1}(t-\tau))$$
$$\times(\alpha_{i,2}\tau + \alpha_{j,2}(t-\tau))d\tau. \tag{6.23}$$

Finally, the covariance and correlation coefficient can be calculated as

$$\text{Cov}_i(X_1(0,t), X_2(0,t))) = E_i(X_1(0,t)X_2(0,t))$$
$$- E_i(X_1(0,t))E_i(X_2(0,t)), \tag{6.24}$$
$$\rho_i(t) = \text{Cov}_i(X_1(0,t), X_2(0,t))$$
$$\times (\sigma_i(X_1(0,t))\sigma_i(X_2(0,t)))^{-1}. \tag{6.25}$$

Now, we suppose that each arrival of the νth flow gives a batch of ν customers, $v = 1, \ldots, k$. Let $Y_i(t)$ be the total number of customers arriving during interval $(0,t)$, if the ith state of the random environment occurs first:

$$Y_i(t) = (X_1(0,t), \dots, X_k(0,t))(1, \dots, k)^T, \quad t \geq 0. \tag{6.26}$$

We wish to calculate the corresponding probability distribution $P\{Y_i(t) = n\}$, $n = 0, 1, \dots$.

Let $\Omega(n, k)$ be a set of k-dimensional integer vectors defined as follows:

$$\Omega(n,k) = \{(\omega_1 \cdots \omega_k)^T : (1, \dots, k)(\omega_1 \cdots \omega_k)^T = n\}. \tag{6.27}$$

Then,

$$\begin{aligned}\{Y_i(t) = n\} = \bigcup_{\omega \in \Omega(n)} \{(X_1(0,t), \dots, X_k(0,t))\\ = (\omega_1 \cdots \omega_k)\}, \quad n = 0, 1, \dots.\end{aligned}$$

The addends of the sum are mutually disjoint events. Therefore, with respect to formulas (6.21) and (6.22), we have

$$P\{Y_i(t) = n\} = \sum_{j=1}^{2} \sum_{\omega \in \Omega(n)} P_{i,j}(\omega, t), \; n = 0, 1, \dots. \tag{6.28}$$

Let us present some sets $\Omega(n, k)$, $k \geq n$:

$$\Omega(1,k) = \left\{ \begin{pmatrix} 1 \\ 0 \\ \cdots \\ 0 \end{pmatrix} \right\}, \quad \Omega(2,k) = \left\{ \begin{pmatrix} 2 \\ 0 \\ 0 \\ \cdots \\ 0 \end{pmatrix}, \begin{pmatrix} 0 \\ 1 \\ 0 \\ \cdots \\ 0 \end{pmatrix} \right\},$$

$$\Omega(4,k) = \left\{ \begin{pmatrix} 3 \\ 0 \\ 0 \\ 0 \\ \cdots \end{pmatrix}, \begin{pmatrix} 1 \\ 1 \\ 0 \\ 0 \\ \cdots \end{pmatrix}, \begin{pmatrix} 0 \\ 0 \\ 1 \\ 0 \\ \cdots \end{pmatrix} \right\},$$

$$\Omega(4,k) = \left\{ \begin{pmatrix} 4 \\ 0 \\ 0 \\ 0 \\ 0 \\ \dots \end{pmatrix}, \begin{pmatrix} 2 \\ 1 \\ 0 \\ 0 \\ 0 \\ \dots \end{pmatrix}, \begin{pmatrix} 0 \\ 2 \\ 0 \\ 0 \\ 0 \\ \dots \end{pmatrix}, \begin{pmatrix} 1 \\ 0 \\ 1 \\ 0 \\ 0 \\ \dots \end{pmatrix}, \begin{pmatrix} 0 \\ 0 \\ 0 \\ 1 \\ 0 \\ \dots \end{pmatrix} \right\}.$$

6.6 Example

Practical calculations assume the substitution of an infinite number of addends with a finite number of addends denoted $n\text{max}$, which is set such that its increase does not change the results essentially.

First, only one Poisson flow was considered. The initial data were fixed as follows:

$$\alpha = \begin{pmatrix} 1 \\ 2 \end{pmatrix}, \quad \lambda = \begin{pmatrix} 0.2 \\ 0.3 \end{pmatrix}.$$

The densities $f(i,j,\tau,t,n\text{max}) = f_{i,j}(\tau,t)$, and $fUn(i,\tau,t, n\text{max}) = f_{i,1}(\tau,t) + f_{i,2}(\tau,t)$ for $n\text{max} = 10$ or 20, and $t = 40$ are presented in Fig. 6.1. The normalization condition for the density $fUn(1,\tau,t,n\text{max})$ was verified for $n\text{max} = 10$ and $t = 5$:

$$\int_0^5 fUn(1,\tau,5,10)d\tau = 0.632, \exp(-\lambda_1 5) = 0.368.$$

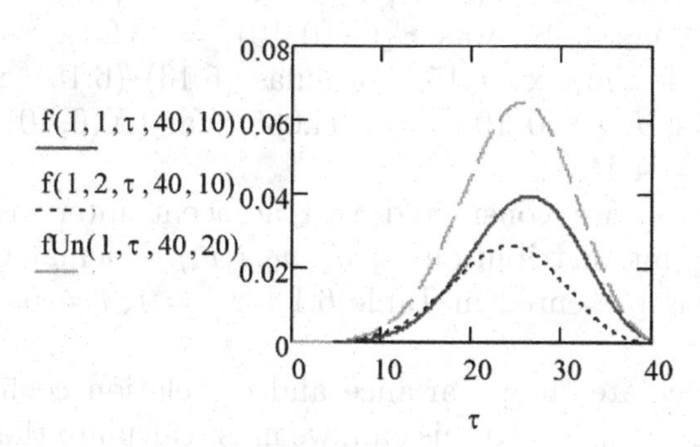

Fig. 6.1. Densities for $n\text{max} = 10$ or 20, and $t = 40$.

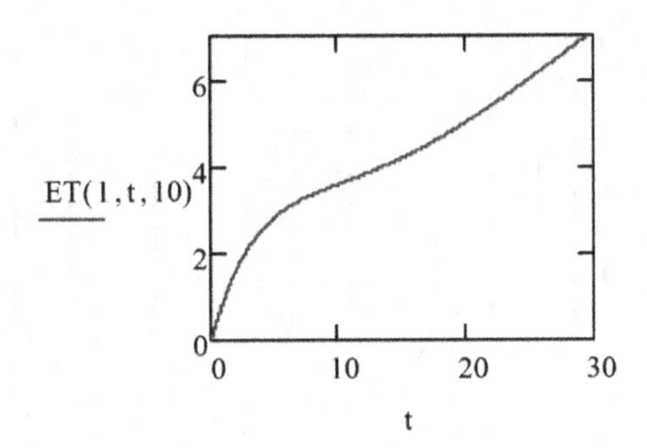

Fig. 6.2. Expectation $E(T_1(t)IJ(0) = 1)$.

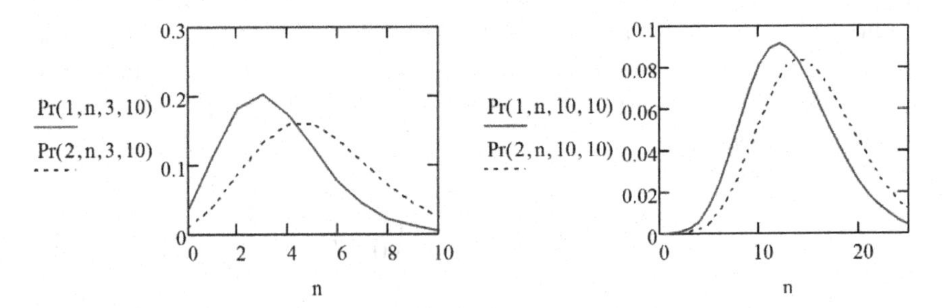

Fig. 6.3. Probability $\Pr_i(n,t)$.

Figure 6.2 depicts the expectation $ET(1,t,10) = E(T_1(t))$ (see formula (6.3)) for $n\max = 10$.

The probability $\Pr_i(n,t)$ (see formula (6.7)) is plotted in Fig. 6.3. Here, $\Pr(i,n,t,n\max) = \Pr_i(n,t)$, $i = 1$ or 2, $t = 3$ or 10.

The expectation $E_i(X(0,t))$, calculated using formula (6.10) for $t = 10$ and $n\max = 10$, was $E_1(X(0,10)) = 13.205$. Similar values are obtained for $n\max = 15$. Formulas (6.13)–(6.15) give the following results: $E_1(X(0,10)^2) = 194.035$, $\mathrm{Var}_1(X(0,10)) = 19.653$, $\sigma_1(X(0,10)) = 4.443$.

Furthermore, we considered two adjacent intervals $(0,t)$ and (t,t^*). The joint distribution $P_i(n_1,n_2,t,t^*)$, calculated using formula (6.16), is presented in Table 6.1 for $i = 1$, $t = 5$, $t^* = 7$, and $n\max = 10$.

Let us calculate the covariance and correlation coefficient using formulas (6.23)–(6.25). To this end, we must calculate the probability $q_i(t)$ that the initial state i occurs at instant t (formula (6.4)). First,

Table 6.1. Joint distribution $P_i(n_1, n_2, 5, 7)$.

$n_1 \| n_2$	0	1	2	3	4	5	6
0	$3.6 \cdot 10^{-4}$	$7.9 \cdot 10^{-4}$	$8.8 \cdot 10^{-4}$	$6.9 \cdot 10^{-4}$	$7.4 \cdot 10^{-4}$	$2.4 \cdot 10^{-4}$	$1.2 \cdot 10^{-4}$
1	$1.9 \cdot 10^{-3}$	$4.1 \cdot 10^{-3}$	$4.6 \cdot 10^{-3}$	$3.7 \cdot 10^{-3}$	$2.3 \cdot 10^{-3}$	$1.3 \cdot 10^{-3}$	$6.4 \cdot 10^{-4}$
2	$4.9 \cdot 10^{-3}$	0.011	0.012	$9.8 \cdot 10^{-3}$	$6.3 \cdot 10^{-3}$	$3.5 \cdot 10^{-3}$	$1.8 \cdot 10^{-3}$
3	$8.7 \cdot 10^{-3}$	0.019	0.022	0.018	0.012	$6.5 \cdot 10^{-3}$	$3.3 \cdot 10^{-3}$
4	0.012	0.026	0.03	0.025	0.016	$9.4 \cdot 10^{-3}$	$4.9 \cdot 10^{-3}$
5	0.013	0.029	0.034	0.028	0.019	0.011	$5.9 \cdot 10^{-3}$
6	0.012	0.027	0.032	0.027	0.019	0.011	$6.1 \cdot 10^{-3}$
7	0.010	0.023	0.028	0.024	0.017	0.010	$5.7 \cdot 10^{-3}$
8	$7.5 \cdot 10^{-3}$	0.017	0.02	0.019	0.014	$8.7 \cdot 10^{-3}$	$4.8 \cdot 10^{-3}$
9	$5.3 \cdot 10^{-3}$	0.012	0.016	0.014	0.010	$6.7 \cdot 10^{-3}$	$3.8 \cdot 10^{-3}$
10	$3.5 \cdot 10^{-3}$	$8.3 \cdot 10^{-3}$	0.011	$9.9 \cdot 10^{-3}$	$7.5 \cdot 10^{-3}$	$4.9 \cdot 10^{-3}$	$2.8 \cdot 10^{-3}$
11	$2.2 \cdot 10^{-3}$	$5.3 \cdot 10^{-3}$	$7.0 \cdot 10^{-3}$	$6.6 \cdot 10^{-3}$	$5.1 \cdot 10^{-3}$	$3.4 \cdot 10^{-3}$	$2.0 \cdot 10^{-3}$
12	$1.3 \cdot 10^{-3}$	$3.3 \cdot 10^{-3}$	$4.3 \cdot 10^{-3}$	$4.2 \cdot 10^{-3}$	$3.3 \cdot 10^{-3}$	$2.2 \cdot 10^{-3}$	$1.3 \cdot 10^{-3}$
13	$7.6 \cdot 10^{-4}$	$1.9 \cdot 10^{-3}$	$2.6 \cdot 10^{-3}$	$2.5 \cdot 10^{-3}$	$2.0 \cdot 10^{-3}$	$1.4 \cdot 10^{-3}$	$8.1 \cdot 10^{-4}$

we consider the following generator:

$$A = \begin{pmatrix} -\lambda_1 & \lambda_1 \\ \lambda_2 & -\lambda_2 \end{pmatrix} = \begin{pmatrix} -0.2 & 0.2 \\ 0.3 & -0.3 \end{pmatrix}.$$

The vector of eigenvalues, matrix of eigenvectors, and the inverse matrix are as follows:

$$\chi = \begin{pmatrix} 0 \\ -0.5 \end{pmatrix}, \; B = \begin{pmatrix} 0.707 & -0.555 \\ 0.707 & 0.832 \end{pmatrix}, \; B^{-1} = \begin{pmatrix} 0.849 & 0.566 \\ -0.721 & 0.721 \end{pmatrix}.$$

The matrix $\Pr(t) = (\Pr_{i,j}(t))$ of transition probabilities during time t is the following:

$$\Pr(t) = \begin{pmatrix} \Pr_{1,1}(t) & \Pr_{1,2}(t) \\ \Pr_{1,1}(t) & \Pr_{2,2}(t) \end{pmatrix} = B \begin{pmatrix} 1 & 0 \\ 0 & \exp(-0.5t) \end{pmatrix} B^{-1}$$
$$= \begin{pmatrix} 0.6 + 0.4 * \exp(-0.57t) & 0.4 - 0.4 * \exp(-0.57t) \\ 0.6 - 0.6 * \exp(-0.57t) & 0.4 + 0.6 * \exp(-0.57t) \end{pmatrix}.$$

Therefore, it is possible to calculate the probability $q_i(t) = \Pr(t)_{i,i}$, obtaining the following results: $q_1(5) = 0.633$, $E_1(X(0{,}5)) = 6.266$, $E_1(X(5{,}7)) = 2.294$, $E_2(X(5{,}7)) = 3.559$,

$E_1(X(0{,}5)X(5{,}7)) = 17.769$, $\sigma_1(X(0{,}5)) = 2.882$, $\mathrm{Var}_1(X(5{,}7) = 2.576$, $\mathrm{Var}_2(X(5{,}7) = 3.943$.

This enables the calculation of the covariance (see formula (6.18)):

$$\begin{aligned}\mathrm{Cov}_1(X(0{,}5), X(5{,}7)) &= E_1(X(0{,}5)X(5{,}7)) \\ &\quad - E_1(X(0{,}5))(q_1(5)E_1(X(5{,}7)) \\ &\quad + (1 - q_1(5))E_2(X(5{,}7))) \\ &= 0.485.\end{aligned}$$

Next, the correlation coefficients are calculated. Formula (6.19) yields $\mathrm{Var}(X(5{,}7|J(0) = 1)) = 3.449$. Therefore,

$$\begin{aligned}\rho_1(X(0{,}5), X(5{,}7)) &= \mathrm{Cov}_i(X(0,t)X(5,7))\sigma_i(X(0,t))^{-1} \\ &\quad \times (\mathrm{Var}(X(5{,}7)|J(0) = 1)))^{-\frac{1}{2}} = 0.091.\end{aligned}$$

Finally, we consider a case of two Poisson flows with the following matrix of intensities:

$$\alpha = \begin{pmatrix} \alpha_{1,1} & \alpha_{1,2} \\ \alpha_{2,1} & \alpha_{2,2} \end{pmatrix} = \begin{pmatrix} 1 & 2 \\ 3 & 1 \end{pmatrix}.$$

Formulas (6.8)–(6.15) and (6.21)–(6.23) give the following results for $i = 1$, $t = 5$:

$$\begin{gathered} E_1(X_1(0{,}5)) = 6.266, \;\; E_1(X_2(0{,}5)) = 12.469, \\ E_1(X(0{,}5)X(0{,}5)) = 74.04, \\ \mathrm{Var}_1(X_1(0{,}5)) = 8.308, \;\; \mathrm{Var}_1(X_2(0{,}0.5)) = 20.638, \\ \sigma_1(X(0{,}0.5)) = 2.882, \;\; \sigma_1(X_2(0{,}0.5)) = 4.543. \end{gathered}$$

Finally, we obtain the following values for the covariance and the correlation coefficient:

$$\mathrm{Cov}_i(X_1(0{,}5), X_2(0{,}5))) = -4.084, \;\; \rho_1(t) = -0.312.$$

Lastly, we consider the case of two Poisson flows ($k = 2$). The arrival of the first flow provides one customer, and the second flow

Table 6.2. Joint distribution $P_i(\omega_1, \omega_2, 1)$.

ω_1	0	1	0	2	1	0	3	2	1	0	4
ω_2	0	0	1	0	1	2	0	1	2	3	0
P	0.225	0.233	0.095	0.123	0.101	0.021	0.044	0.055	0.023	0.003	0.012
ω_1	3	2	1	0	5	4	3	2	1	0	Σ
ω_2	1	2	3	4	0	1	2	3	4	5	—
P	0.021	0.013	0.004	0.000	0.003	0.006	0.006	0.002	0.001	0.000	0.992

Table 6.3. Probabilities $P\{Y_i(t) = n\}$.

$t \setminus n$	0	1	2	3	4	5
0.25	0.699	0.178	0.095	0.021	0.006	0.001
0.50	0.482	0.248	0.165	0.064	0.027	0.009
0.75	0.330	0.256	0.205	0.110	0.057	0.025
1.00	0.225	0.233	0.218	0.145	0.088	0.047
1.25	0.153	0.198	0.211	0.166	0.116	0.071
1.50	0.104	0.161	0.193	0.173	0.135	0.093
1.75	0.070	0.128	0.169	0.169	0.146	0.111
2.00	0.048	0.099	0.143	0.158	0.149	0.123
2.25	0.032	0.075	0.119	0.143	0.145	0.129
$t \setminus n$	6	7	8	9	10	Σ
0.25	0.000	0.000	0.000	0.000	0.000	1.000
0.50	0.003	0.001	0.000	0.000	0.000	0.999
0.75	0.010	0.004	0.002	0.001	0.000	0.999
1.00	0.023	0.011	0.005	0.002	0.001	0.998
1.25	0.041	0.022	0.11	0.006	0.003	0.997
1.50	0.060	0.036	0.021	0.011	0.006	0.993
1.75	0.078	0.051	0.032	0.019	0.011	0.984
2.00	0.094	0.066	0.045	0.029	0.018	0.972
2.25	0.105	0.080	0.058	0.040	0.027	0.953

provides two customers, where

$$\Omega(n,k)= \begin{cases} \left\{ \binom{n-2\eta}{\eta} : \eta = 0, 1, \ldots, n/2 \right\}, & \text{if } n \text{ is even number,} \\ \left\{ \binom{n-2\eta}{\eta} : \eta = 0, 1, \ldots, (n-1)/2 \right\}, & \text{if } n \text{ is odd number.} \end{cases}$$

The parameters of the distribution of sojourn times in both states are the following: $\lambda_1 = 0.2$ and $\lambda_2 = 0.3$.

The intensity $\alpha_{i,\nu}$ of the νth flow if the ith state occurs is determined by the matrix:

$$\alpha = \begin{pmatrix} \alpha_{1,1} & \alpha_{1,2} \\ \alpha_{2,1} & \alpha_{2,2} \end{pmatrix} = \begin{pmatrix} 1 & 0.4 \\ 2 & 1 \end{pmatrix}.$$

Table 6.2 lists the joint distribution of the number of arrivals for the two flows during time $t = 1$, calculated using formulas (6.21) and (6.22).

Table 6.3 lists the probabilities $P\{Y_i(t) = n\}$ of customer numbers for different times t calculated using formula (6.28).

The value in the row $t = 1$, can be obtained from Table 6.2.

6.7 Estimation of parameters

We confine our study to the case of a single Poisson flow. It is assumed that parameters λ_1 and λ_2 of the alternating MC $J(t)$ are known, but the intensities of the flows α_1 and α_2 are not. We wish to estimate those having the following statistical data for each observation ν: $t(\nu)$ is observation duration, $i(\nu)$ is the initial state of MC $J(t)$, and $x(\nu)$ is the number of arrivals within interval $(0, t(\nu))$.

The moment method is the simplest method for point estimation [9,10]. The entire sample is divided into two non-empty subsamples for the two initial states of the MC $J(t)$. Let t_1, $E(T_1)$, and X_1 be the total observation time, expectation of sojourn time in state 1, and number of arrivals for the first subsample, respectively, and t_2, $E(T_2)$, and X_2 the same parameters for the second subsample. The values t_1, t_2, X_1, and X_2 are calculated using the given subsamples, whereas expectations $E(T_1)$ and $E(T_2)$ are calculated using formula (6.10). The following procedure is applied to the νth observation:

(1) The values $i(\nu)$, $t(\nu)$, and $x(\nu)$ are fixed.
(2) The expectation of arrival $E_{i(\nu)}(X(0, t(\nu)))$ is calculated by formula (6.10).
(3) The values $x(\nu)$ are added for the fixed $i(\nu) = 1$ or $i(\nu) = 2$. Let $X_i, i = 1, 2$, be the corresponding sums.

The moment method yields the following linear equation system for estimating parameters α_1 and α_2:

$$\tilde{\alpha}_1 E(T_1) + \tilde{\alpha}_2(t_1 - E(T_1)) = X_1,$$
$$\tilde{\alpha}_1 E(t_2 - E(T_2)) + \tilde{\alpha}_2 E(T_2) = X_2.$$

By setting

$$\tilde{\alpha} = \begin{pmatrix} \tilde{\alpha}_1 \\ \tilde{\alpha}_2 \end{pmatrix}, \quad X = \begin{pmatrix} X_1 \\ X_2 \end{pmatrix}, \quad A = \begin{pmatrix} E(T_1) & t_1 - E(T_1) \\ t_2 - E(T_2) & E(T_2) \end{pmatrix},$$

we have the following estimate:

$$\tilde{\alpha} = A^{-1}X. \tag{6.29}$$

6.8 Simulation study

We now verify the parameter estimation procedure as described earlier. To this end, we conducted a simulation [11]. Assuming that the parameters have the following values:

$$\alpha = \begin{pmatrix} 1 \\ 2 \end{pmatrix}, \quad \lambda = \begin{pmatrix} 0.2 \\ 0.3 \end{pmatrix}.$$

Parameters $\lambda_1 = 0.2$ and $\lambda_2 = 0.3$ are known. We wish to estimate the unknown intensities of the flows α_1 and α_2.

We divide the entire sample into two subsamples with the same number of observations. The first and second subsamples contain, respectively, observations under the first initial state $i(\nu) = 1$, and the second initial state $i(\nu) = 2$ of the MC $J(t)$. The duration of all observations $t(\nu)$ is constant and equal to t.

The simulation was used to investigate how the sample size influenced the accuracy of the results. The operations conducted for each observation ν includes the following:

(1) The alternating process with initial state $i(\nu)$ is simulated during time t, by means of sequential generation of exponentially distributed variables with parameters λ_1 or λ_2. The sojourn time $\tau_i(\nu)$ in state $i(\nu)$ is fixed.

(2) A random number $x(\nu)$ is generated with respect to the Poisson distribution with parameter $\alpha_i\tau_i(\nu)+\alpha_{not(i)}(t-\tau_i(\nu))$. The values $x(\nu)$ are the result of the given observation.

A reiteration of this procedure provides the sum X_i for the considered subsample. The same procedure is repeated for each subsample. Now, we can estimate the unknown parameters using formula (6.29).

The simulation was organized as follows. First, we sequentially considered small samples of size m without exceeding 10–20 observations. The estimates were averaged to obtain one realization, and the procedure was repeated to get various realizations.

Table 6.4 summarizes the estimates $\tilde{\alpha}_1$ (upper rows) and $\tilde{\alpha}_2$ (lower rows) for the following initial data. The duration of each observation t was 15 min. The sample size m was 20, including 10 observations under the first initial state and 10 observations under the second initial state. Each sample comprised five subsamples. The last row is the average of all the realizations. Each realization included the averaging of 1, 2 to 10 estimates.

We see that the averages of all $20 \times 10 \times 5 = 1{,}000$ observations (see the last two rows in the last column) are 0.944 for $\tilde{\alpha}_1$ and 1.884

Table 6.4. Statistical analysis with small samples.

Realiz. Number	1	2	3	4	5	6	7	8	9	10
1	0.515	0.635	0.554	0.643	0.689	0.723	0.707	0.666	0.730	0.812
	3.121	2.762	1.842	1.984	2.056	1.713	1.792	1.957	1.903	1.787
2	0.318	0.972	0.941	0.903	1.082	1.063	1.003	1.041	1.029	0.997
	3.304	1.887	2.003	2.045	1.716	1.725	1.869	1.815	1.851	1.852
3	0.974	0.855	0.699	0.864	0.750	0.809	0.920	0.937	0.924	0.886
	0.000	1.190	1.853	1.722	1.037	2.049	1.838	1.859	1.897	2.025
4	0.541	0.657	0.845	0.906	0.843	0.876	0.887	0.846	0856	0.933
	2.642	2.638	2.114	1.586	1.841	1.854	1.803	1.917	1.983	1.892
5	0.732	0.671	0.988	1.163	1.004	0.992	1.031	1.025	1.008	1.093
	2.736	2.717	2.074	1.797	2.120	2.157	2.052	2.020	2.033	1.865
Average	0.616	0.758	0.805	0.896	0.874	0.893	0.910	0.903	0.909	0.944
	2.361	2.239	1.997	1.827	1.954	1.900	1.871	1.914	1.933	1.884

Table 6.5. Statistical analysis with large samples.

Realiz. Number	1	2	3	4	5	6	7	8	9	10
1	1.261	0.983	0.952	1.134	1.041	0.877	0.940	0.804	0.909	0.932
	1.573	2.010	2.054	1.767	1.032	2.210	2.105	2.206	2.166	2.087
2	0.276	0.744	0.773	0.810	0.894	0.894	0.924	0.987	0.595	0.973
	2.971	2.369	2.377	2.329	2.225	2.225	2.160	2.071	2.143	2.085
3	1.288	1.204	1.181	1.195	0.975	1.034	1.017	1.035	1.016	1.023
	1.481	1.632	1.737	1.680	2.104	1.975	2.002	1.978	2.034	2.011
4	0.800	0.878	0.962	1.046	1.073	0.981	1.033	1.062	1.079	1.022
	2.255	2.142	2.202	1.883	1.818	1.995	1.926	1.854	1.808	1.909
5	1.838	1.088	1.163	1.165	1.128	1.072	0.960	1.012	1.098	1.046
	0.793	1.834	1.718	1.679	1.795	1.919	2.080	1.981	1.819	1.900
Average	1.093	0.979	1.006	1.070	1.022	0.972	0.975	0.998	1.012	0.999
	1.815	1.997	1.981	1.868	1.975	2.065	2.055	2.018	1.994	1.998

for $\tilde{\alpha}_2$, instead of 1 and 2, respectively. This implies that our estimates have some bias for small samples.

We also considered large sample sizes. We began with $m = 100$ and sequentially increased the size to 1,000 observations with the step 100. This corresponds to one realization. The simulation results are listed in Table 6.5, whose structure is analogous to that of Table 6.4, except that the sample size corresponds to the number of each column multiplied by 100 (or multiplied by m).

The two estimates in the last rows of the last column, which correspond to $100 \times 10 \times 5 = 5{,}000$ observations, are $\tilde{\alpha}_1 = 0.999$ and $\tilde{\alpha}_2 = 1.998$. This implies that our estimates are consistent.

6.9 Conclusion

Recently, MMPPs have attracted significant attention from researchers. In this chapter, new results have been obtained for alternating Poisson processes. The following properties of these flows were investigated: distribution of the number of arrivals at a given interval, and covariance and correlation coefficients of the number of arrivals for two adjacent intervals. We also considered many Poisson flows along with an estimation procedure.

The proposed model can be used in modern telecommunication and data-processing systems to process calls or data packages in a random external environment.

Appendix

Lemma A1. *Let X, Y, and Z be independent random variables, with Z corresponding to a Boolean variable, $P\{Z = 1\} = q$, and X and Y having expectation $E(X)$ and $E(Y)$, variances* $\mathrm{Var}(X)$ *and* $\mathrm{Var}(Y)$, *and distributions densities $h_X(t)$ and $h_Y(t)$, respectively. Then, the variance of random variable $S = ZX + (1 - Z)Y$ is given by*

$$\mathrm{Var}(S) = q\mathrm{Var}(X) + (1-q)\mathrm{Var}(Y) + q(1-q)(E(X) - E(Y))^2.$$

Proof. Because $Z^2 = Z$, $E(Z^2) = q$, $E(Z(1-Z)) = 0$, then

$$\begin{aligned}
\mathrm{Var}(S) &= \mathrm{Var}(ZX + (1-Z)Y) \\
&= E((ZX + (1-Z)Y)^2)(E(ZX + (1-Z)Y))^2 \\
&= E((ZX)^2) + E(((1-Z)Y)^2) + 2E(ZX(1-Z)Y) \\
&\quad - (E(ZX + (1-Z)Y))^2 \\
&= qE(X^2) + (1-q)E(Y^2) - (qE(X) + (1-q)E(Y))^2 \\
&= qE(X^2) \mp qE(X)^2 + (1-q)E(Y^2) \mp (1-q)E(Y)^2 \\
&\quad - ((qE(X))^2 + ((1-q)E(Y))^2 + 2qE(X)(1-q)E(Y)) \\
&= (qE(X^2) - qE(X)^2) + ((1-q)E(Y^2) - (1-q)E(Y)^2) \\
&\quad + (qE(X)^2 - (qE(X))^2) + ((1-q)E(Y)^2 \\
&\quad - ((1-q)E(Y))^2) - 2q(1-q)E(X)E(Y) \\
&= q\mathrm{Var}(X) + (1-q)\mathrm{Var}(Y) + q(1-q)(E(X) - E(Y))^2.
\end{aligned}$$

Hence, the lemma is proven. Formula (6.19) can be derived by setting $Z = 1$, *if* $J(t) = i$, and $Z = 0$, otherwise, $X = X(t, t^*)$ with the condition that $J(t) = i$, and $Y = X(t, t^*)$ with the condition that $J(t) \neq i$. □

References

[1] Gnedenko, B.V., Belyaev, Yu.K., Solovyev, A.D., *Mathematical Methods of Reliability.* Academic Press, New York, 1969.

[2] Kijima, M., *Markov Processes for Stochastic Modeling.* The University Press, Cambridge, UK, 1997.

[3] Ross, Sh.M., *Applied Probability Models with Optimization Applications.* Dover Publications, Inc., New York, 1992.

[4] Du, Q., A monotonicity result for a single-server queue subject to a Markov-modulated Poisson process. *Journal of Applied Probability,* 1995;32(4):1103–1111.

[5] Fischer, W., Meier-Hellstern, K., The Markov modulated Poisson process cookbook. *Performance Evaluation,* 1992;18:149–171.

[6] Özekici, S., Soyer, R., Semi-Markov modulated Poisson process, probabilistic and statistical analysis. *Mathematical Methods of Operational Research,* 2006;64:125–144.

[7] Andronov, A.M., Vishnevsky, V.M., Markov-modulated continuous time finite Markov chain as the model of hybrid wireless communication channels operation. *Automatic Control and Computer Sciences,* 2016;50(3):125–132.

[8] Vishnevsky, V.M., Andronov, A.M., Estimating the throughput of wireless hybrid systems operating in a semi-Markov stochastic environment. *Automation and Remote Control,* 2017;78(12):2154–2165.

[9] Cramer, H., *Mathematical Methods of Statistics.* Princeton University Press, New York, 1946.

[10] Rao, C.R., *Linear Statistical Inference and Its Applications.* John Wiley & Sons, Inc., New York, London, Sydney, 1973.

[11] Gentle, J.E., *Elements of Computational Statistics.* Springer-Verlag, New York, Berlin, Heidelberg, 2002.

Chapter 7

Analysis of Markov-Modulated Poisson Processes

7.1 Introduction

Markov-modulated Poisson processes were discussed in Section 2.2. As described earlier, the process operates in a random environment consisting of a finite continuous-time irreducible Markov chain $J(t)$, $t \geq 0$ (see Chapter 1). The chain is defined by nonnegative parameters $\lambda_{i,j} \geq 0$, $i, j = 1, \ldots, k$. The sojourn time in state i is assumed to follow an exponential distribution with parameter $\Lambda_i = \sum_{j=1}^{k} \lambda_{i,j}$. When the sojourn time in one state comes to an end, the chain transfers in the state j with probability $\lambda_{i,j}/\Lambda_i$. If the Markov chain is under the ith state, $i = 1, \ldots, k$, the Poisson process has intensity μ_i.

The generation function for the probability of arrival at time t was presented in Section 2.2. A partial case with the two states ($k = 2$) was considered in Chapter 6.

Now we consider the general case by using the afore-described approach (see Chapter 2). This approach is based on computational calculations. This distinguishes our approach from other typical approaches [1–4].

The density function of the time until the first arrival of the Poisson process is discussed in the following section.

Our second aim was to determine the distribution of the number of arrivals of the Poisson flow during time t. We call *j-arrival* the arrival that occurs when the Markov chain has the jth state at instant t.

We wish to determine the distribution of the number of *j-arrivals* during time $t, j = 1, \ldots, k$.

An interesting application of this process is an inventory problem, which will be also considered.

7.2 Time distribution until the first arrival

Let the Markov chain have initial state i and $f_{i,j}(t), i, j \in \{0, \ldots, k-1\}; t \geq 0$, be the density function of the time until the first arrival jointly with a probability that the jth state of the Markov chain occurs at this instant. An expression for this density can be obtained by considering the probability $\hat{P}_{i,j}(t)$ that the Markov chain $X(t)$ transfers from the ith state to the jth state during time t without arrivals of Poisson flow.

Considering the first jump after initial time 0, the usual reasoning for a small time step $\Delta t > 0$ leads to the following system of equations:

$$\hat{P}_{i,j}(t + \Delta t) = (1 - \Delta t(\Lambda_i + \mu_{\mathrm{i}}))\hat{P}_{i,j}(t) + \sum_{\nu=1}^{k} \lambda_{i,v} \Delta \hat{P}_{v,j}(t) + o(\Delta t), \quad t \geq 0, \ \forall i, j.$$

Furthermore, the equation

$$\frac{1}{\Delta}(\hat{P}_{i,j}(t + \Delta) - \hat{P}_{i,j}(t)) = -(\Lambda_i + \mu_{\mathrm{i}})\hat{P}_{i,j}(t) + \sum_{\nu=1}^{k} \lambda_{i,v} \hat{P}_{v,j}(t) + \frac{o(\Delta t)}{\Delta t}, \quad t \geq 0, \ \forall i, j,$$

leads us to the following system of differential equations with constant coefficients:

$$\dot{\hat{P}}_{i,j}(t) = -(\Lambda_i + \mu_{\mathrm{i}})\hat{P}_{i,j}(t) + \sum_{\nu=1}^{k} \lambda_{i,v} \hat{P}_{v,j}(t), \quad t \geq 0, \ \forall i, j. \quad (7.1)$$

Next, introducing matrices $\hat{P}(t) = (\hat{P}_{i,j}(t))_{k \times k}$ and $\lambda = (\lambda_{i,j})_{k \times k}$, vector $\tilde{\Lambda} = ((\Lambda_1 + \mu_1), \ldots, (\Lambda_k + \mu_{\mathrm{k}}))^T$, and diagonal matrix $\mathrm{diag}(\tilde{\Lambda})$

with vector $\tilde{\Lambda}$ as the main diagonal, we can rewrite system (7.1) in matrix form as

$$\dot{\hat{P}}(t) = -\text{diag}(\tilde{\Lambda})\hat{P}(t) + \lambda\hat{P}(t),$$

$$\dot{\hat{P}}(t) = (\lambda - \text{diag}(\tilde{\Lambda}))\hat{P}(t), \quad t \geq 0. \tag{7.2}$$

This system is known as the *backward Kolmogorov equation.* An explicit solution for this system is known. The matrix $G = \lambda - \text{diag}(\tilde{\Lambda})$ is called *a generator.* Eigenvalues and eigenvectors of this matrix are denoted $\chi_1, \chi_2, \ldots, \chi_k$ and $\beta_1, \beta_2, \ldots, \beta_k$, respectively. In our case all values $\chi_1, \chi_2, \ldots, \chi_k$ are different. Matrix, whose columns are eigenvectors, is denoted $B = (\beta_1, \ldots, \beta_k)$, and its inverse matrix with rows $\tilde{\beta}_1, \ldots, \tilde{\beta}_k$ is denoted $B^{-1} = \tilde{B} = (\tilde{\beta}_1^T, \ldots, \tilde{\beta}_k^T)^T$.

Then

$$\hat{P}(t) = \sum_{i=1}^{k} \exp(\chi_i t)\beta_i\widetilde{\beta}_i, \quad t \geq 0. \tag{7.3}$$

Note, that the first arrival takes place in a small interval $(t, t+\Delta)$, when the Markov chain has the jth state, with probability $\hat{P}_{i,j}(t)\Delta\mu_j$. Therefore,

$$f_{i,j}(t) = \mu_j\hat{P}_{i,j}(t), \quad i, j = 1, \ldots k; \; t \geq 0. \tag{7.4}$$

Now, we can calculate all moments of time T until the first arrival. If the Markov chain has the initial state i, the expectation $E_{i,j}(T)$ jointly with the probability that the jth state of the Markov chain occurs at this instant is

$$E_{i,j}(T) = E(T\delta_j(T)|J(0) = i) = \int_0^\infty tf_{i,j}(t)dt, \quad i, j = 1, \ldots, k, \tag{7.5}$$

where $\delta_j(t)$ is an indicator:

$$\delta_{j(t)} = \begin{bmatrix} 1 & \text{if } J(t) = j, \\ 0, & \text{otherwise.} \end{bmatrix} \tag{7.6}$$

Furthermore, we set

$$E_i(T) = \sum_{j=1}^{k} E_{i,j}(T), \quad i = 1, \ldots k. \tag{7.7}$$

The second moment and variance for the initial state i are calculated as follows:

$$E_i(T^2) = \sum_{j=1}^{k} \int_0^\infty t^2 f_{i,j}(t)dt, \quad i,j = 1,\ldots,k, \tag{7.8}$$

$$\mathrm{Var}_i(T) = E_i(T^2) - E_i(T)^2. \tag{7.9}$$

With the first of our goals achieved, the covariance for two adjacent intervals between arrivals can be calculated.

Let $T^{(2)}$ be the time between the first and second arrivals. Then, for $v = 1, 2, \ldots$:

$$\begin{aligned} E_i((T^{(2)})^v) &= E((T^{(2)})^v | J(0) = i) \\ &= \sum_{j=1}^{k} \int_0^\infty f_{i,j}(\tau)d\tau \sum_{\nu=1}^{k} \int_0^\infty t^v f_{j,\nu}(t)dt, \end{aligned} \tag{7.10}$$

$$\mathrm{Var}_i(T^{(2)}) = \mathrm{Var}(T^{(2)} | J(0) = i) = E_i((T^{(2)})^2) - (E_i(T^{(2)}))^2. \tag{7.11}$$

Further

$$E(TT^{(2)} | J(0) = i) = \sum_{j=1}^{k} E_{i,j}(T) \sum_{\nu=1}^{k} E_{j,\nu}(T). \tag{7.12}$$

Finally we have the following expression for the covariance between T and $T^{(2)}$:

$$\begin{aligned} &\mathrm{Cov}(T, T^{(2)} | J(0) = i) \\ &\quad = E((T - E_i(T))(T^{(2)} - E_i(T^{(2)})) | J(0) = i) \\ &\quad = E(TT^{(2)} - TE_i(T^{(2)}) - E_i(T)T^{(2)} + E_i(T)E_i(T^{(2)}) | J(0) = i) \\ &\quad = E(TT^{(2)} | J(0) = i) - E_i(T)E_i(T^{(2)}). \end{aligned} \tag{7.13}$$

This allows for the calculation of a correlation coefficient between T and $T^{(2)}$, if the Markov chain has an initial state i:

$$\rho_i(T, T^{(2)}) = \frac{\mathrm{Cov}(T, T^{(2)} | J(0) = i)}{\sqrt{\mathrm{Var}_i(T)}\sqrt{\mathrm{Var}_i(T^{(2)})}}. \tag{7.14}$$

7.3 Distribution of time until arrivals

We denote $T^{(n)}$ as the time until the nth arrival. The corresponding cumulative distribution function for the ith initial state denoted as

$$F_i(n,t) = P\{T^{(n)} \le t | J(0) = i\}, \quad t \ge 0, \; n = 1, 2, \ldots$$

Because all eigenvalues $\chi_1, \chi_2, \ldots, \chi_k$ of the generator are non-zero, then

$$F_i(1,t) = \sum_{j=1}^{k} \int_0^t f_{i,j}(\tau)d\tau = \sum_{j=1}^{k} \int_0^t \sum_{\eta=1}^{k} \mu_j \exp(\chi_\eta \tau)\beta_{\eta,i}\widetilde{\beta_{\eta,j}}d\tau$$

$$= \sum_{j=1}^{k} \mu_j \sum_{\eta=1}^{k} \frac{1}{\chi_\eta}(\exp(\chi_\eta t) - 1)\beta_{\eta,i}\widetilde{\beta_{\eta,j}}, \tag{7.15}$$

where $\beta_{\eta,i}$ and $\widetilde{\beta_{\eta,j}}$ are the corresponding elements of the vector-column β_η and vector-row $\widetilde{\beta_\eta}$.

The following expression applies to other values of n:

$$F_i(n,t) = \sum_{j=1}^{k} \int_0^t f_{i,j}(\tau)F_j(n-1, t-\tau)d\tau, \quad n = 2, 3, \ldots \tag{7.16}$$

For example, for $n = 2$:

$$F_i(2,t) = \sum_{j=1}^{k} \int_0^t f_{i,j}(\tau)F_j(1, t-\tau)d\tau$$

$$= \sum_{j=1}^{k} \int_0^t \sum_{\eta=1}^{k} \mu_j \exp(\chi_\eta \tau)\beta_{\eta,i}\widetilde{\beta_{\eta,j}}$$

$$\times \sum_{\theta=1}^{k} \mu_\theta \sum_{\varsigma=1}^{k} \frac{1}{\chi_\varsigma}(\exp(\chi_\varsigma(t-\tau)) - 1)\beta_{\varsigma,j}\widetilde{\beta_{\varsigma,,\theta}}d\tau$$

$$= \sum_{j=1}^{k}\sum_{\eta=1}^{k} \mu_j \beta_{\eta,i}\widetilde{\beta_{\eta,j}} \sum_{\theta=1}^{k} \mu_\theta \sum_{\varsigma=1}^{k} \frac{1}{\chi_\varsigma}\beta_{\varsigma,j}\widetilde{\beta_{\varsigma,,\theta}}\exp(\chi_\varsigma t))$$

$$\times \int_0^t (\exp((\chi_\eta - \chi_\varsigma)\tau) - \exp(\chi_\eta \tau))d\tau$$

$$= \sum_{j=1}^{k} \sum_{\eta=1}^{k} \mu_j \beta_{\eta,i} \widetilde{\beta_{\eta,j}} \sum_{\theta=1}^{k} \mu_\theta \sum_{\varsigma=1,\varsigma\neq\eta}^{k} \frac{1}{\chi_\varsigma} \beta_{\varsigma,j} \widetilde{\beta_{\varsigma,,\theta}} \exp(\chi_\varsigma t))$$

$$\times \left[\frac{1}{\chi_\eta - \chi_\varsigma} \{\exp((\chi_\eta - \chi_\varsigma)t) - 1\} - \frac{1}{\chi_\eta} \{\exp(\chi_\eta t) - 1\} \right]$$

$$+ \sum_{j=1}^{k} \sum_{\eta=1}^{k} \mu_j \beta_{\eta,i} \widetilde{\beta_{\eta,j}} \sum_{\theta=1}^{k} \mu_\theta \frac{1}{\chi_\eta} \beta_{\eta,j} \widetilde{\beta_{\eta,,\theta}} \exp(\chi_\eta t))$$

$$\times \left(t - \frac{1}{\chi_\eta} (\exp(\chi_\eta \tau) - 1) \right).$$

As it can be observed, the numerical realization of formula (7.16) is complicated; therefore, a specific approach is presented below in Section 7.5.

7.4 Distribution of the number of arrivals of the Poisson flow

We denote the number of arrivals until the time t as $N(t)$, and its cumulative distribution function, if the ith state of the Markov chain occurs initially, as

$$Q_i(n,t) = P\{N(t) = n | J(0) = i\}, \quad t \geq 0, \ n = 0, 1, \ldots$$

Obviously, for $t > 0$:

$$\begin{aligned} Q_i(0,t) &= 1 - F_i(1,t), \\ Q_i(n,t) &= F_i(n,t) - F_i(n+1,t), \quad n = 1, 2, \ldots \end{aligned} \tag{7.17}$$

The disadvantage of this expression is that it depends on the repeated convolution of the distribution function, which requires specific numerical procedures. These procedures are described in detail below. However, two initial moments of the random variable $N(t)$ are simply calculated. Let $P_{i,j}(t)$ be the probability that the Markov chain is in the jth state at instant t (without a Poisson flow of arrivals) if the initial state is equal to i. This probability and the corresponding matrix $P(t)$ are calculated as demonstrated earlier, setting $\mu_i = 0$, $i = 1, \dots, k$.

Consequently, the expectation is calculated as follows:

$$\begin{aligned} H(t)_i &= E(N(t)|J(0) = i) \\ &= \sum_{j=1}^{k} \int_0^t \mu_j P_{i,j}(\tau) d\tau, \quad t \geq 0, \ i = 1, \dots, k. \end{aligned} \tag{7.18}$$

The expression for the second moment and $i = 1, \dots, k$, is more complex:

$$\begin{aligned} E(N(t)^2|J(0) = i) = 2\sum_{j=1}^{k} \int_0^t \mu_j P_{i,j}(\tau) \sum_{l=1}^{k} \int_0^{t-\tau} \mu_l P_{j,l}(\theta) d\theta d\tau \\ - \int_0^t \mu_i P_{i,i}(\tau) \int_0^{t-\tau} \mu_i P_{i,i}(\theta) d\theta d\tau, \quad t \geq 0. \end{aligned} \tag{7.19}$$

The last two equations can be used to verify the numerical results with respect to the calculation of the arrival distribution. The corresponding aspects are considered in detail.

7.5 Computational aspects

Due to the multiple convolutions required, the computational cost increases as the number of arrivals increases. Consequently, we use a special approach that will be described for the nth arrival.

Let us represent the distribution function $F_i(n, t)$, $t \geq 0$, $n = 3, 4, \dots$, for the nth arrival time through a two-dimensional matrix $M(n) = (M_{i,\eta}(n))$, of dimensions $k \times n\text{max}$. The ith row of the matrix contains data for the initial state with number i. Continuous values of t are replaced with the points of the axis $(0, t\text{max})$ with the interval $\Delta > 0$. Supposing that $t\text{max} = \Delta \times (n\text{max} - 1)$, the value $M_{i,\eta}(n)$ is defined as follows:

$$M_{i,\eta}(n) = F_i(n, (\eta - 1)\Delta), \quad \eta = 1, \dots, n\text{max}. \tag{7.20}$$

Matrices M are stored in the computer's memory and used instead of function $F_i(n, t)$. A necessary value $F_i(n, t)$ for $t \in (\Delta(\eta - 1), \ \Delta\eta)$

is presented as follows:

$$
\begin{aligned}
Ft_i(M(n),t) &= \frac{1}{\Delta}((\Delta\eta - t)M_{i,\eta}(n) + (t - \Delta(\eta-1))M_{i,\eta+1}(n)) \\
&= \left(\eta - \frac{t}{\eta}\right) M_{i,\eta}(n) + \left(\frac{t}{\eta} - \eta + 1\right) M_{i,\eta+1}(n).
\end{aligned}
\tag{7.21}
$$

Now we can calculate the function $F_i(n,t)$ using the function $Ft_i(M(n-1), t-\tau)$ instead of the function $F_i(n-1,t)$ in the formula (7.16).

The matrices for the previous and the next steps, $M(n-1)$ and $M(n)$, $n \in \{2, 3, \ldots\}$, are also stored in the computer's memory. The densities $\{f_{i,j}(t)\}$ used in formula (7.5) are calculated using formula (7.4). The distribution functions $\{F_i(1,t)\}$ for the first step are calculated using formula (7.15).

7.6 Stationary distribution

Now, we consider the stationary distribution of a Markov-modulated Poisson flow. This implies that the initial state of the Markov chain at the beginning of the considered time interval has a stationary distribution. Let π_i be the stationary probability of the ith state, $i = 1, \ldots, k$. Evidently, the stationary value of the index in question is an average of its non-stationary values with weights $\{\pi_i\}$.

The vector of stationary probabilities $\pi = (\pi_1 \quad \cdots \quad \pi_k)^T$ can be obtained by setting $\mu_i = 0$, $i = 1, \ldots k$, in formulas (7.1)–(7.3). The generator is defined for this purpose as $G = \lambda - \text{diag}(\Lambda)$ where Λ is used instead of $\tilde{\Lambda}$. This case is considered in Section 1.2 (see Theorem 1.2, and formula (1.12)). Let B be a matrix, whose columns are eigenvectors $B = (\beta_1, \ldots, \beta_k)$ and whose inverse matrix is $B^{-1} = \tilde{B} = (\tilde{\beta}_1^T, \ldots, \tilde{\beta}_k^T)^T$, and i^* be the number of zero eigenvalues. Then

$$
\lim_{t\to\infty} P(t) = \beta_{i*}\widetilde{\beta_{i*}},
$$

Since all the components of vector β_{i*} are the same, all rows of matrix $\beta_{i*}\widetilde{\beta_{i*}}$ are also the same. Each such row gives a vector of the stationary probabilities of states $\pi = (\pi_1 \quad \cdots \quad \pi_k)^T$:

$$\beta_{i*}\widetilde{\beta_{i*}} = \begin{pmatrix} \pi \\ \cdots \\ \pi \end{pmatrix}.$$

We denote the indices of interest for the stationary case by adding power s to the index used earlier, e.g., $f_j^s(t)$ instead of $f_{i,j}(t)$, and omitting the initial state. Then

$$f_j^s(t) = \sum_{i=1}^{k} \pi_i f_{i,j}(t), \quad j = 1, \ldots, k; \ t \geq 0. \tag{7.22}$$

The expectation of $N(t)$ arrivals during time t is of significant interest. For the stationary case, we use Eq. (7.14):

$$\begin{aligned} E^s(N(t)) &= \sum_{i=1}^{k} \pi_i E(N(t)|J(0) = i) \\ &= \sum_{i=1}^{k} \pi_i \sum_{j=1}^{k} \int_0^t \mu_j P_{i,j}(\tau) d\tau = \int_0^t \sum_{j=1}^{k} \sum_{i=1}^{k} \pi_i P_{i,j}(\tau) \mu_j d\tau \\ &= \int_0^t \sum_{j=1}^{k} \pi P(t)^{\langle j \rangle} \mu_j \, d\tau = \int_0^t \sum_{j=1}^{k} \pi_j \mu_j \, d\tau = t \sum_{j=1}^{k} \pi_j \mu_j. \end{aligned} \tag{7.23}$$

The average intensity of the Markov-modulated Poisson flow is

$$\mu^s = \sum_{j=1}^{k} \pi_j \mu_j$$

The resulting expectation is

$$E^s(N(t)) = \mu^s t \tag{7.24}$$

and we have strong proof.

7.7 Inventory problem

We consider the so-called *single-period inventory model with stochastic demand* [5,6]. The last is presented by the Markov-modulated

Poisson process considered above. The intensity of demands equals μ_i if the external random environment has the ith state, $i = 1, \ldots, k$. Each demand requests a one store unit. The initial storage consists of s units. The cost of each unit is denoted as c. The profit per unit used is d. The time period considered is equal to t. It is assumed that the initial state i of the Markov chain is known.

The reward as a function of the initial store s and time t, is calculated as follows:

$$R(s,t) = d\widetilde{N}(s,t) - cs,$$

where $\widetilde{N}(s,t)$ is a number of used units of the supply during time t.

It is necessary to find an initial store s with maximal expected reward.

The total demand $N(t)$ during time t follows the distribution $Q_i(n,t)$ given by formula (7.17). Because

$$\widetilde{N}(s,t) = \min\{s, N(t)\},$$

the expected reward $R(s,t)$ for $s \geq 0$ is calculated as follows:

$$\begin{aligned} E_i(R(s,t)) &= E(R(s,t)|J(0) = i) = dE(\widetilde{N}(s,t)) - cs \\ &= d\sum_{n=1}^{s} nQ_i(n,t) + ds\left(1 - \sum_{n=0}^{s} Q_i(n,t)\right) - cs. \end{aligned} \tag{7.25}$$

The optimal value of the initial store s is determined by enumeration.

If the initial state follows a stationary distribution $\pi = (\pi_1 \quad \cdots \quad \pi_k)^T$, then the corresponding expected reward can be expressed as follows:

$$E^s(R(s,t)) = \sum_{i=1}^{k} \pi_i E_i(R(s,t)). \tag{7.26}$$

7.8 Example

Our example includes the following initial conditions. The number of Markov chain states is $k = 3$; the intensities of the transitions between the states are

$$\lambda = \begin{pmatrix} 0 & 0.4 & 1.0 \\ 0.7 & 0 & 1.1 \\ 0 & 1.5 & 0 \end{pmatrix};$$

and the intensities of arrivals for different states are

$$\mu = (1 \quad 2 \quad 3)^T.$$

Therefore,

$$\Lambda = \begin{pmatrix} \Lambda_1 \\ \Lambda_2 \\ \Lambda_3 \end{pmatrix} = \begin{pmatrix} 0 & 0.4 & 1.0 \\ 0.7 & 0 & 1.1 \\ 0 & 1.5 & 0 \end{pmatrix} \begin{pmatrix} 1 \\ 1 \\ 1 \end{pmatrix} = \begin{pmatrix} 1.4 \\ 1.8 \\ 1.5 \end{pmatrix},$$

$$\tilde{\Lambda} = ((\Lambda_1 + \mu_1), \ldots, (\Lambda_k + \mu_k))^T = (2.4 \quad 3.8 \quad 4.5)^T.$$

The generator and its associated matrices and vectors are

$$G = \lambda - \mathrm{diag}(\tilde{\Lambda}) = \begin{pmatrix} -2.4 & 0.4 & 1.0 \\ 0.7 & -3.8 & 1.1 \\ 0 & 1.5 & -4.5 \end{pmatrix},$$

$$\chi = \begin{pmatrix} \chi_1 \\ \chi_2 \\ \chi_3 \end{pmatrix} = \begin{pmatrix} -1.875 \\ -3.449 \\ -5.376 \end{pmatrix},$$

$$B = \begin{pmatrix} -0.849 & 0.707 & -0.217 \\ -0.459 & -0.406 & -0.492 \\ -0.262 & -0.579 & 0.843 \end{pmatrix}, \quad B^{-1} = \begin{pmatrix} -0.727 & -0.545 & -0.506 \\ 0.598 & -0.896 & -0.369 \\ 0.185 & -0.785 & 0.775 \end{pmatrix}$$

Now we can use formulas (7.3) and (7.4) for calculations. Figure 7.1 contains the graph of the time density until the first arrival $f(t, 1, j) = f_{1,j}(t)$ calculated by formula (7.4).

Some of the computed results are presented below. Table 7.1 contains values $E_{i,j}(T)$ for $i, j = 1, 2, 3$. Table 7.2 lists the values calculated using Eqs. (7.8)–(7.14). We observe that the adjacent intervals between arrivals T and $T2$ are weakly correlated.

Figure 7.2 depicts the cumulative distribution functions $PFirst(t)_1 = F_1(1, t)$, $PSec(t)_1 = F_1(2, t)$, and $PThree(t)_1 = F_1(3, t)$ derived from formula (7.16).

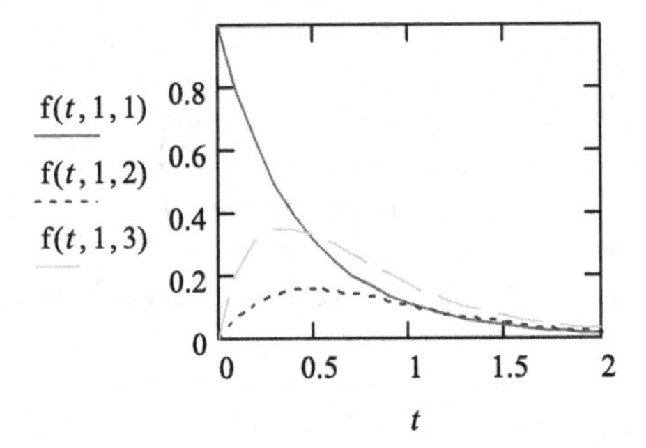

Fig. 7.1. Graph of the density $f_{1,j}(t)$.

Table 7.1. Expectation $E_{i,j}(T)$.

	$i = 1$	$j = 2$	$j = 3$	Σ
$i = 1$	0.209	0.167	0.281	0.957
$i = 2$	0.071	0.229	0.195	0.495
$i = 3$	0.030	0.122	0.234	0.386

Table 7.2. Values associated with density $f\ (t, i, j)$.

	$i = 1$	$i = 2$	$i = 5$
$E(T^2 \| J(0) = i)$	0.773	0.501	0.339
$Var(T \| J(0) = i)$	0.340	0.256	0.189
$E_i(T^{(2)})$	0.527	0.478	0.417
$E_i((T^{(2)})^2)$	0.562	0.479	0.386
$\text{Var}(T^{(2)} \| J(0) = i)$	0.284	0.250	0.211
$\text{Cov}(T, T^{(2)} \| J(0) = i)$	−0.018	−0001	0.010
$\rho_i(T, T^{(2)})$	−0.057	−0.005	0.048

Because the calculation of $F_i(n,t)$ for $n > 3$ requires extended computer processing time, the approach described in Section 7.5 is used. Figure 7.3 shows that the obtained results $\text{PThree}(t)_1$ and $Ft_1(M(3), t)$ closely coincide for $n = 3$. This allowed us to use the proposed approach for $n > 3$. The corresponding distributions for $M = M(3)$, $M(4)$, $M(5)$, $M(6)$, $M(7)$ are presented in Fig. 7.4.

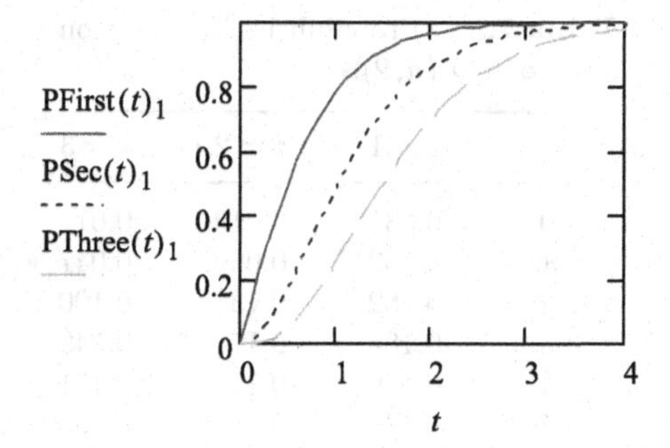

Fig. 7.2. Distribution function $F_1(n, t)$ for $n = 1, 2, 3$.

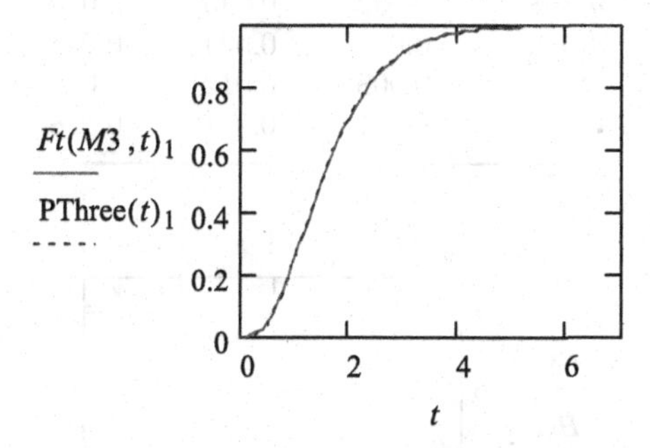

Fig. 7.3. Distribution function $Ft_1(M(3), t)$ and $PThree(t)_1$.

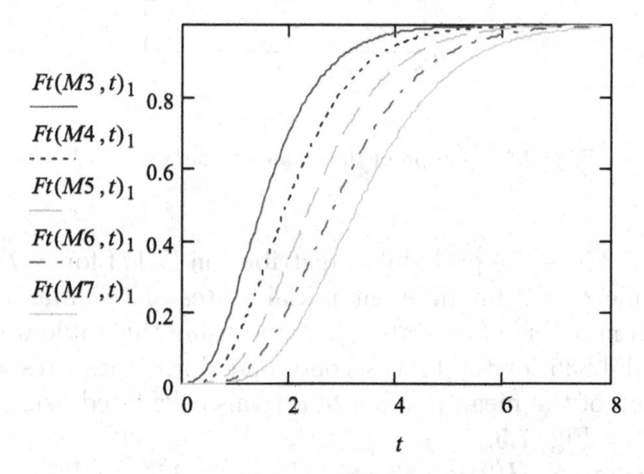

Fig. 7.4. Distribution functions $Ft_1(M, t)$.

Table 7.3. Probabilities distribution functions $Q_i(n, 2)$.

	$i = 1$	$i = 2$	$i = 3$
$n = 0$	0.035	0.019	0.01
$n = 1$	0.102	0.069	0.047
$n = 2$	0.162	0.131	0.100
$n = 3$	0.185	0.172	0.149
$n = 4$	0.170	0.177	0.171
$n = 5$	0.133	0.153	0.163
$n = 6$	0.091	0.113	0.132
$n = 7$	0.057	0.074	0.094
$n = 8$	0.032	0.045	0.060
$n = 9$	0.018	0.024	0.036
$n = 10$	0.008	0.012	0.018
$n = 11$	0.007	0.011	0.018

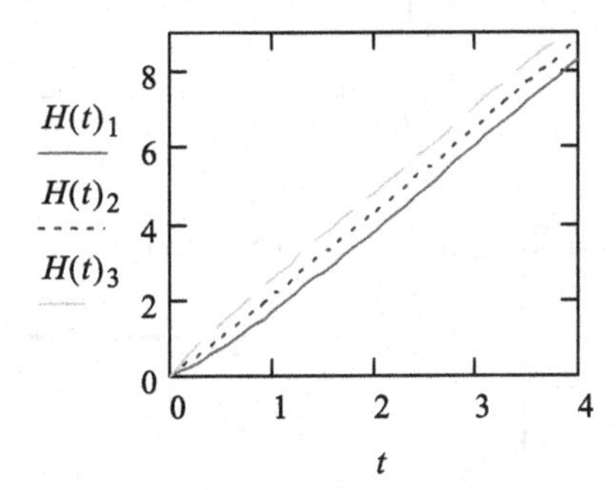

Fig. 7.5. Graph of the mean number of arrivals.

Table 7.3 lists the probability distribution (7.17) for $N(t)$ arrivals during time $t = 2$ for different initial states of the Markov chain. The mean numbers of arrivals calculated using this table were 3.846, 4.333, and 4.835 for the first, second, and third states, respectively.

A graph of the mean number of arrivals calculated using (7.18) is presented in Fig. 7.5.

For example, $H(2)_1 = 3.834$, $H(2)_2 = 4.326$, $H(3)_1 = 4.835$. The difference with respect to the previous results is insignificant, thus demonstrating the accuracy of the computational procedures described in Section 7.5.

We now consider the case where the Markov chain has a stationary distribution initially. Setting $\mu_i = 0$, $i = 1, 2, 3$, we have the following results:

$$\chi = \begin{pmatrix} \chi_1 \\ \chi_2 \\ \chi_3 \end{pmatrix} = \begin{pmatrix} -2.714 \\ -1.986 \\ 0 \end{pmatrix},$$

$$B = \begin{pmatrix} -0.371 & 0.816 & 0.577 \\ -0.584 & 0.178 & 0.577 \\ 0.732 & -0.55 & 0.577 \end{pmatrix}, \quad B^{-1} = \begin{pmatrix} 0.736 & -1.382 & 0.646 \\ 1.321 & -0.106 & -0.215 \\ 0.337 & 0.675 & 0.720 \end{pmatrix}.$$

We see that the third eigenvalue χ_3 equals zero, therefore $i^* = 3$. The product of the third column β_{i*} of matrix B and the third row $\widetilde{\beta_{i*}}$ of matrix B^{-1} is

$$\beta_{i*}\widetilde{\beta_{i*}} = \begin{pmatrix} 0.577 \\ 0.577 \\ 0.577 \end{pmatrix} (0.337 \quad 0.675 \quad 0.720) = \begin{pmatrix} 0.195 & 0.390 & 0.416 \\ 0.195 & 0.390 & 0.416 \\ 0.195 & 0.390 & 0.416 \end{pmatrix}.$$

Also, the vector of stationary probabilities of the Markov chain is

$$\pi = (\pi_1 \quad \pi_2 \quad \pi_3)^T = (0.195 \quad 0.390 \quad 0.416).$$

Now, we can calculate the stationary values of all indices of interest, particularly the stationary intensity of the arrivals:

$$\mu^s = \sum_{j=1}^{k} \pi_j \mu_j = 0.195 \times 1 + 0.390 \times 2 + 0.416 \times 3 = 2.223.$$

The average stationary number of arrivals during time t is calculated by formula (7.24):

$$E^s(N(t)) = \mu^s t = 2.223t.$$

We conclude this example by considering an inventory model. Let the cost and profit of each unit be $c = 3$ and $d = 6$, respectively. The expected reward (7.25) as a function of the initial stores s and state i is presented in Table 7.4 for $t = 2$. The data in the last row corresponds to the stationary case.

The maximum expected reward is marked in bold style. The optimal values of the initial stores correspond to these values.

Table 7.4. Expected reward $E_i(R(s,2))$.

s	1	2	3	4	5	6	7	8	9
$i = 1$	2.79	4.97	6.17	**6.27**	5.35	3.62	1.36	−1.25	−4.06
$i = 2$	2.88	5.36	7.04	**7.70**	7.29	5.96	3.96	1.51	−1.21
$i = 3$	2.93	5.58	7.64	8.79	**8.93**	8.08	6.44	4.24	1.68
E^s	2.89	5.38	7.13	**7.88**	7.60	6.39	4.49	2.11	−0.56

7.9 Discussion

A question that may naturally arise for the readers of this chapter is: Why is the method presented in Section 1.2 not used? Let us attempt to answer this.

We consider the two-dimensional continuous-time Markov chain $Z(t) = (J(t), X(t))$, where $J(t) \in \{0, \ldots, k-1\}$ is a state of the random environment, $X(t) \in \{0, 1, \ldots, m\}$ is the number of arrivals of the Poisson flow, $t \geq 0$, $X(0) = 0$, and parameter m is the maximum number of considered arrivals.

Let us recall that transition intensities between states of a random environment are $\lambda_{i,j} \geq 0$, $i, j = 0, \ldots, k-1$, and that the Poisson flow has intensity μ_i if the random environment is in the ith state.

Let $g_{(i,x),(j,y)}$ be the transfer intensity from state (i, x) to state (j, y) of the Markov chain $Z(t)$, $i, j \in \{0, \ldots, k-1\}$; $x, y \in \{0, 1, \ldots, m\}$.

For non-zero intensities, we have

$$g_{(i,x),(j,x)} = \lambda_{i,j}, \quad x = 0, \ldots, m,$$

$$g_{(i,x),(i,x+1)} = \mu_i, \quad x = 0, \ldots, m-1,$$

where $i, j \in \{0, \ldots, k-1\}$.

To use the approach described in Section 1.2, we must transfer from a two-dimensional process $Z(t) = (J(t), X(t))$ to a one-dimensional chain $Y(t)$. State (i, x) of process $Z(t)$ corresponds to state $n(i, x) = (i + kx)$ of the chain $Y(t)$. The inverse transfer is described using the following two operators:

$$i(n) = \text{mod}\,(n, k),$$

$$x(n) = \frac{1}{k}(n - \text{mod}\,(n, k)).$$

Note that

$$n(i, x+1) = (i + (k+1)x) = (i + kx) + x = n(i, x) + x,$$
$$n(j, x) = (j + kx) = (i + kx) + (j - i) = n(i, x) + (j - i).$$

Thus, the non-zero intensities presented above can be rewritten as follows:

$$g_{n(i,x),n(i,x)+(j-i)} = \lambda_{i,j}, \quad x = 0, \ldots m,$$
$$g_{n(i,x),n(i,x)+x} = \mu_i, \quad x = 0, \ldots m-1.$$

Now we can use the proposed approach to calculate the generator G, its vector of eigenvalues χ, and the matrix of eigenvectors B. Let us apply this procedure to the data considered in Example 1.

First, we present the correspondence between the unary (n) and binary (i, x) number of states, as shown in Table 7.5.

Generator G, up to the ninth column, is presented in Table 7.6. Let us remark that its rank equals 11. It is normally because the sum

Table 7.5. Correspondence between the unary (n) and binary (i, x) numbers.

n	0	1	2	3	4	5	6	7	8	9	10	11
i	0	1	2	0	1	2	0	1	2	0	1	2
x	0	0	0	1	1	1	2	2	2	3	3	3

Table 7.6. Generator G.

Gen(3) =

	0	1	2	3	4	5	6	7	8	9
0	−2.4	0.4	1	1	0	0	0	0	0	0
1	0.7	−3.8	1.1	0	2	0	0	0	0	0
2	0	1.5	−4.5	0	0	3	0	0	0	0
3	0	0	0	−2.4	0.4	1	1	0	0	0
4	0	0	0	0.7	−3.8	1.1	0	2	0	0
5	0	0	0	0	1.5	−4.5	0	0	3	0
6	0	0	0	0	0	0	−2.4	0.4	1	1
7	0	0	0	0	0	0	0.7	−3.8	1.1	0
8	0	0	0	0	0	0	0	1.5	−4.5	0
9	0	0	0	0	0	0	0	0	0	−1.4
10	0	0	0	0	0	0	0	0	0	0.7
11	0	0	0	0	0	0	0	0	0	...

Table 7.7. Matrix of eigenvectors B.

$B(3) =$

	0	1	2	3	4	5	6	7
0	−0.849	0.707	−0.217	0.849	−0.707	0.217	−0.849	0.707
1	−0.459	−0.406	−0.492	0.459	0.406	0.492	−0.459	−0.406
2	−0.262	−0.579	0.843	0.262	0.579	−0.843	−0.262	−0.579
3	0	0	0	0	0	0	0	0
4	0	0	0	0	0	0	0	0
5	0	0	0	0	0	0	0	0
6	0	0	0	0	0	0	0	0
7	0	0	0	0	0	0	0	0
8	0	0	0	0	0	0	0	0
9	0	0	0	0	0	0	0	0
10	0	0	0	0	0	0	0	0
11	0	0	0	0	0	0	0	...

of all columns yields the zero vector. Therefore, any column equals the sum of the other negative columns.

The vector of eigenvalues χ is the following:

$$\chi(3) = (-1.875 \quad -3.449 \quad -5.376 \quad -1.875 \quad -3.449 \quad -5.376 \quad \ldots)^T.$$

It can be observed that values −1.875, −3.449, −5.376 repeat periodically. Therefore, a case of multiple eigenvalues occurs, and the previous approach does not work. It can be clearly seen, considered the matrix B of the eigenvectors. This is presented till the seventh column in Table 7.7.

The rank of this matrix is 3; therefore, its inverse matrix does not exist, and the main formula (7.3) cannot be used.

7.10 Conclusion

We now complete our consideration of the Markov-modulated Poisson processes with the following conclusions. First of all, the main results of the previous chapter are generalized to many states of the random environment. Secondly, the proposed approach can be applied to various modifications of Markov-modulated Poisson processes.

References

[1] Fischer, W., Meier-Hellstern, K.S., The Markov-modulated Poisson process (MMPP) cookbook. *Perform Evaluation*, 1993;18(2):149–171.

[2] London, J., Ozekici, S., Soyer, R., Markov-modulated Poisson model for software reliability. *European Journal of Operation Research*, 2013; 229:404–410.

[3] Özekici, S., Soyer, R., Semi-Markov modulated Poisson process, Probabilistic and statistical analysis. *Mathematical Methods Operation Reserve*, 2006;64:125–144.

[4] Pacheco, A., Tang, L.C., Prabhu, N.U., *Markov-Modulated Processes & Semiregenerative Phenomena*. World Scientific, New Jersey, London, 2009.

[5] Ozekici, S., Parlar, M., Inventory models with unreliable suppliers in a random environment. *Annals of Operations Research*, 1999;91: 123–136.

[6] Prabhu, N.U., *Queues and Inventories. A Study of Their Basic Stochastic Processes*. John Wiley & Sons, Inc., New York, London, Sydney, 1965.

Chapter 8

Monitoring of Continuous-Time Markov Chains

8.1 Introduction

Monitoring is the process of observing an object in order to control its state. It is of great significance for controlling various technological processes [1–3].

Often, indicators of interest cannot be observed directly, and it is necessary to use other indicators that are connected to those of interest. The main probabilistic processes $X(t)$ discussed in this book are immersed in an external random environment, represented as a continuous-time finite Markov chain $J(t)$. We now consider a situation involving two processes $X(t)$ or $J(t)$ in which only one process is fully observed and recognizing the state of the other is desired.

Such monitoring of process $X(t)$ is investigated simply because process $J(t)$ is determined as primary, whereas process $X(t)$ is modulated with respect to the obtained realization of process $J(t)$. The aforementioned methods can also be used in such situations. We consider the *Markov-modulated Poisson process* as an example.

A continuous-time finite Markov chain $J(t)$ has k states. The chain is assumed to be fully observed. In this environment, the Poisson flow $X(t)$ operates with a variable intensity. The flow has intensity α_i, $i = 1, \ldots, k$, if the ith state of chain $J(t)$ occurs. We are interested in determining how many arrivals occur within interval $(0, t)$. As the chain is observed, we know sojourn times $\tau_1, \ldots, \tau_k$ of the chain in each state. In this case, the random variable $X(t)$ follows a Poisson distribution with parameters $\Lambda = \alpha_1\tau_1 + \cdots + \alpha_{k1}\tau_k$. Therefore, the distribution of flow states at time t is as follows:

$$P\{X(t) = n\} = \frac{1}{n!}\Lambda^n \exp(-\Lambda), \quad n = 0, 1, \ldots \tag{8.1}$$

By contrast, when the state of process $J(t)$ is recognized based on process $X(t)$, it is a serious problem about a *Hidden Markov chain*. Numerous studies have been conducted on this problem [3–10]. We discuss one in the current chapter. An alternating Poisson process is considered first, followed by a more general Markov-modulated Poisson process.

8.2 Alternating Poisson process

Let us assume that $J(t)$ is an alternating Markov chain with two alternating states, 1 and 2. The sojourn times in states 1 and 2 are independent random variables that follow exponential distributions with known parameters λ_1 and λ_2. The inbound Poisson flow $X(t)$ has known intensities of α_1 and α_2 for the first and second states, respectively. We aim to recognize the state of process $J(t)$ at instant t based on the information available from process $X(t)$.

The solution to this problem depends on the type of information obtained. First, we consider the case where the state of process $X(t)$, that is, the number of Poisson flow arrivals, is used.

The following information is at our disposal: the initial state of the Markov chain $J(0) = i$ and the values $X(0) = 0$ and $X(t) = n$ at instants 0 and t.

The solution follows from Bayes formula for $i, j = 1, 2$; $t \geq 0$:

$$P\{J(t) = j|J(0) = i, X(t) = n\} = \frac{P\{J(t) = j, X(t) = n|J(0) = i\}}{P\{X(t) = n|J(0) = i\}}. \tag{8.2}$$

The probabilities appearing on the right-hand side of formula (8.2) are calculated using formulas (6.5)–(6.7):

$$\begin{aligned} P\{J(t) = j, X(t) = n|J(0) = i\} &= P_{i,j}(n, t), \\ P\{X(t) = n|J(0) = i\} &= P_{i,1}(n, t) + P_{i,2}(n, t). \end{aligned} \tag{8.3}$$

It would be interesting to consider how the results change if additional information is available. In addition, we know that at instant $t^* < t$ n^* arrivals occur, $n^* < n$. In this case,

$$\begin{aligned}
&P\{J(t*) = j * |J(0) = i, X(t*) = n*\}\\
&\quad = \frac{P\{J(t) = j*, X(t*) = n * |J(0) = i\}}{P\{X(t*) = n * |J(0) = i\}},\\
&P\{J(t) = j, J(t*) = j * |J(0) = i, X(t*) = n*, X(t) = n\}\\
&\quad = P\{J(t*) = j * |J(0) = i, X(t*) = n*\}\\
&\quad \times \frac{P\{J(t-t*) = j, X(t-t*) = n - n * |J(0) = j*\}}{P\{X(t-t*) = n - n * |J(0) = j*\}},
\end{aligned}$$

where all probabilities are calculated using formulas (6.5)–(6.7).

Finally,

$$\begin{aligned}
&P\{J(t) = j|J(0) = i, X(t*) = n*, X(t) = n\}\\
&\quad = P\{J(t) = j, J(t*) = 1|J(0) = i, X(t*) = n*, X(t) = n\}\\
&\quad + P\{J(t) = j, J(t*) = 2|J(0) = i, X(t*) = n*, X(t) = n\}.
\end{aligned} \tag{8.4}$$

Naturally, if state j yields the maximal probability (8.4) the state should be set to $J(t) = j$.

Note that this result can be generalized for lots of intermediate observations.

The case of a Markov-modulated Poisson flow in which the number of Markov chain states is greater than two is considered next. This is a more difficult case because there is no explicit expression for the distribution of the number of arrivals. Therefore, it is impossible to apply the Bayes' formula. Instead of probabilities, we use the average number of arrivals and consider how the results obtained using formula (8.2) change.

The expectation $E_{i,j}(X(0,t))$ of the number of arrivals jointly with the probability that the last state is j, is calculated using formulas (6.8) and (6.9). The conditional expectation $E_i(X(0,t)|J(t) = j)$ with the conditions $J(0) = i$ and $J(t) = j$ is calculated as follows:

$$E_i(X(0,t)|J(t) = j) = \frac{E_{i,j}(X(0,t))}{P\{J(t) = j|J(0) = i\}}. \tag{8.5}$$

Probability $P\{J(t) = j|J(0) = i\}$ can be calculated in two ways. First, as

$$P\{J(t) = j|J(0) = i\} = \int_0^t f_{i,j}(\tau, t)d\tau, \tag{8.6}$$

where $f_{i,j}(\tau, t)$ is the density of the sojourn time $T_i(t)$ at point τ for the initial state i during time t jointly with the probability that the final state equals j (see formulas (6.1) and (6.2)).

The second option is as follows. Let $\mathrm{Pr}(t) = (\mathrm{Pr}_{i,j}(t))$ be a corresponding 2×2 matrix. In this case, we have the following *generator* (see Section 1.2):

$$A = \begin{pmatrix} -\lambda_1 & \lambda_1 \\ \lambda_2 & -\lambda_2 \end{pmatrix}.$$

Let $\chi = (\chi_1 \quad \chi_2)^{\mathrm{T}}$ be the vector of eigenvalues and $B = (\beta_1 \quad \beta_2)$ the matrix of eigenvectors of A. Then,

$$\mathrm{Pr}(t) = B \,\mathrm{diag}\big(\exp(\chi t)\big) B^{-1}. \tag{8.7}$$

Definition 8.1. Let $X(0, t) = n$ and:

$$\Delta_j = |n - E_i(X(0, t)|J(t) = j)|, \quad j = 1, 2. \tag{8.8}$$

If j^* is such that $\Delta_{j^*} = \min\{\Delta_1, \Delta_2\}$, then j^* is called pseudo-Bayes recognizer (estimator) of $J(t)$.

Finally, instead of expectations, a generation function can be used. Let $\psi_{i,j}(z, t)$ be the generation function of arrivals $X(t)$ jointly with the probability that the last state is j if the initial state is i. Using the indicator $\delta_j(t)$ of the random event $\{J(t) = j\}$, we can rewrite

$$\psi_{i,j}(z, t) = E(z^{X(t)}\delta_j(t)|J(0) = i), \quad i, j = 1,\, 2; \;\; z, t \geq 0.$$

The generation function of argument $z \geq 0$ for a Poisson distribution with parameters a is $\exp(-a(1-z))$. It follows from (6.1) and (6.2) that

$$\psi_{i,j}(z, t) = \int_0^t f_{i,j}(\tau, t) \exp(-(\alpha_i \tau + \alpha_{not(i)}(t - \tau))(1 - z))d\tau, \quad i \neq j, \tag{8.9}$$

$$\psi_{i,i}(z, t) = \int_0^t f_{i,i}(\tau, t) \exp(-(\alpha_i \tau + \alpha_{not(i)}(t - \tau))(1 - z))d\tau + \exp(-\lambda_i) \exp(-\alpha_i t(1 - z)). \tag{8.10}$$

The conditional generation function given $J(0) = i$ and $J(t) = j$ is defined as

$$\psi_i(z, j, t) = E(z^{X(t)}|J(0) = i, J(t) = j), \quad i, j = 1, 2; \;\; z, t \geq 0.$$

Analogously with (8.5), we have

$$\psi_i(z, j, t) = \frac{\psi_{i,j}(z, t)}{P\{J(t) = j|J(0) = i\}}. \tag{8.11}$$

If n is the value of the observed arrival number $X(0, t)$ then the empirical generation function is z^n. For example, the measure of its concordance with the generation function in (8.11) is

$$\tilde{\Delta}_j = \int_{t*}^{t} |\psi_i(z, j, t) - z^n) dz, \tag{8.12}$$

where t^* is the same value from interval $(0,\ t)$.

The value $j \in \{1, 2\}$ for which function (8.12) has a minimum value, is assumed *as a recognized value* of $J(t)$.

8.3 Example

Our example has the following input data: $\lambda_1 = 0$, $\lambda_2 = 0.3$, $\alpha_1 = 1$, and $\alpha_2 = 2$. Figure 8.1 shows the results of $Bayes(i, t, n, j, n\max) = P\{J(t) = j|J(0) = i,\ X(t) = n\}$ as a function of the number of arrivals n for $t = 15$, $n\max = 20$. The probability of the first state is higher if the arrival number n does not exceed 18.

The plot also indicates that the probability of state $J(15)$ is not significantly dependent on its initial state. However, the situation changes when t is small (Fig. 8.2).

Now the question arises: how does the posterior probability (9.2) of the state $J(t)$ change due to the additional information on the number of arrivals $X(t)$ as compared to the probability $\mathrm{Pr}_{i,j}(t)$ of state $J(t)$? This probability can be calculated by formulas (8.6) or (8.7).

The results are given in Figs. 8.3 and 8.4, where $\mathrm{Pr}S(1, 1, \mathrm{t}, 20) = \mathrm{Pr}_{1,1}(t)$. We can see that this distinction is essential.

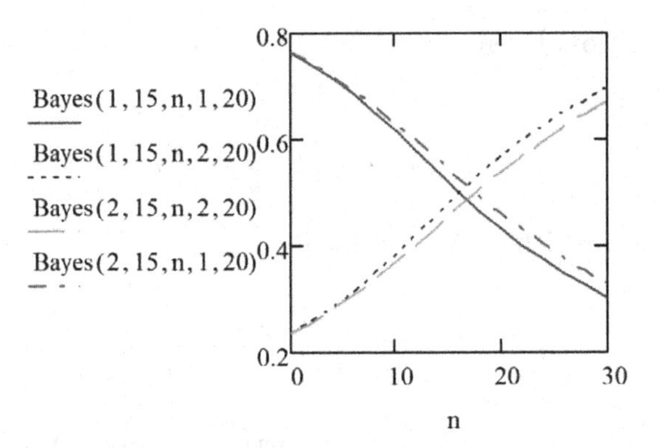

Fig. 8.1. Bayes probabilities (8.2) for $t = 15$.

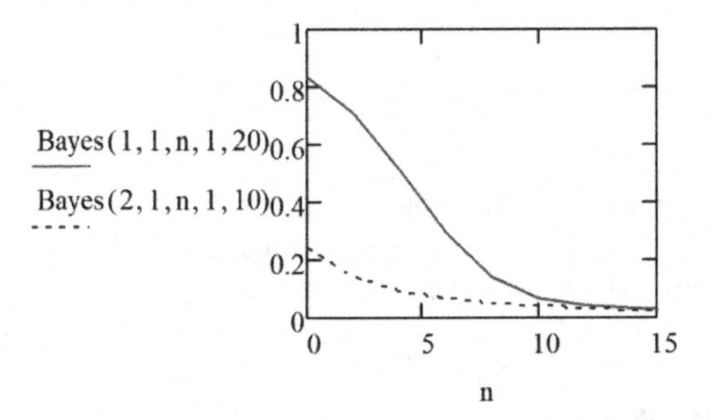

Fig. 8.2. Bayes probabilities (8.2) for $t = 1$.

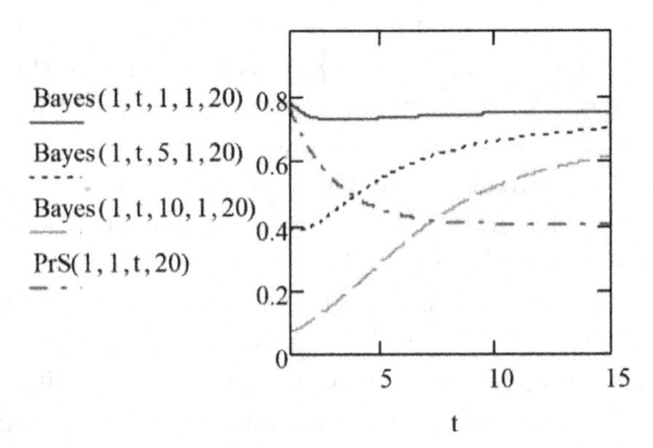

Fig. 8.3. Bayes probabilities Bayes$(i, t, n, j, n\text{max})$ for $i = j = 1$, and $n = 1, 5, 10$, and probability $\text{Pr}_{i,j}(t)$.

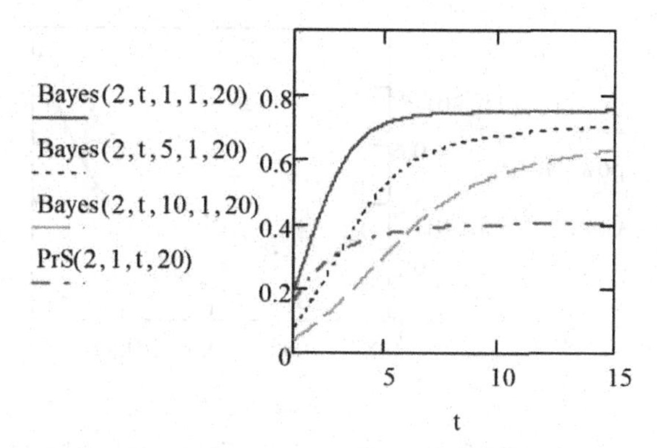

Fig. 8.4. Bayes probabilities Bayes$(i, t, n, j, n\text{max})$ for $i = 2$, $j = 1$, and $n = 1, 5, 10$, and probability $\text{Pr}_{i,j}(t)$.

These examples demonstrate that the Bayes approach yields satisfactory results. It is based on the distribution of the probability $P_{i,j}(n, t)$ that n arrivals occur within the interval $(0,\ t)$, and the final state is j if we initially have state i. Now, we consider how the results change if a different approach is adopted. This allows us to select an approach that could be used if the previously remembered distribution was unknown.

First, we consider the conditional expectation $E_i(X(0,t)|J(t)=j)$ that is calculated by formula (8.5). The recognition criterion Δ_j is calculated by formula (8.8). The value j, giving the smallest Δ_j, is adopted as state $J(t)$. Tables 8.1 and 8.2 summarize the results for the considered example, $t = 15$, $i = 1, 2$, and various numbers of arrivals n.

Table 8.1. Adopted value j of $J(15)$ for $J(0) = 1$.

n	20	21	22	23	24	25	26	27
j	1	1	1	2	2	2	2	2

Table 8.2. Adopted value j of $J(15)$ for $J(0) = 2$.

n	20	21	22	23	24	25	26	27
j	1	1	1	1	1	2	2	2

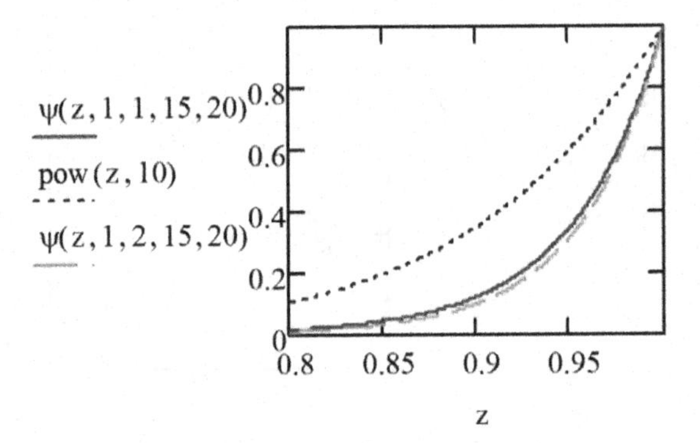

Fig. 8.5. Generation functions for the initial state $i = 1$ and $n = 10$.

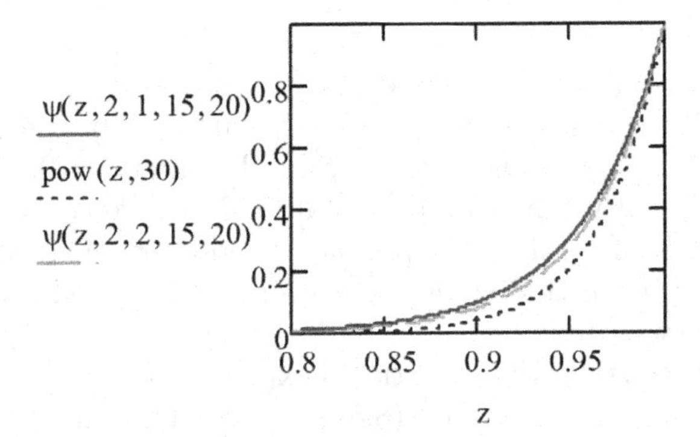

Fig. 8.6. Generation functions for the initial state $i = 2$ and $n = 30$.

We can see that the switching points are $n = 23$ for $J(0) = 1$ and $n = 25$ for $J(0) = 2$. The Bayes approach yielded $n = 17$. Therefore, this distinction is essential.

Second, we consider the conditional generation function $\psi(z, j, t)$, given by formula (8.11). The calculation results are shown in Figs. 8.5 and 8.6, along with function $pow(z, n) = z^n$. The conditional generation function $\psi_i(z, j, t)$ is shown here as $\psi(z, i, j, t, n\max)$, where $n\max$ is the number of addends, which are taken into account in infinite sum.

The recognition criterion $\tilde{\Delta}_j$ is calculated through formula. (8.12) with $t^* = 0$, and $t = 15$. The results are presented in Tables 8.3 and 8.4.

Table 8.3. Recognition criterion $\tilde{\Delta}_j$ for $J(0) = 1$ and $t = 15$.

n	$\tilde{\Delta}_1$	$\tilde{\Delta}_2$
16	0.011	0.016
17	8.142×10^{-3}	0.012
18	5.392×10^{-3}	9.401×10^{-3}
19	3.212×10^{-3}	6.814×10^{-3}
20	1.956×10^{-3}	4.550×10^{-3}
21	2.240×10^{-3}	2.717×10^{-3}
22	4.032×10^{-3}	1.595×10^{-3}
23	5.843×10^{-3}	1.760×10^{-3}
24	7.510×10^{-3}	3.257×10^{-3}
25	9.048×10^{-3}	4.795×10^{-3}

Table 8.4. Recognition criterion $\tilde{\Delta}_j$ for $J(0) = 2$ and $t = 15$.

n	$\tilde{\Delta}_1$	$\tilde{\Delta}_2$
18	9.401×10^{-3}	0.013
19	6.814×10^{-3}	0.010
20	4.550×10^{-3}	7.914×10^{-3}
21	2.717×10^{-3}	5.778×10^{-3}
22	1.595×10^{-3}	3.899×10^{-3}
23	1.760×10^{-3}	2.338×10^{-3}
24	3.257×10^{-3}	1.318×10^{-3}
25	4.795×10^{-3}	1.414×10^{-3}
26	6.220×10^{-3}	2.685×10^{-3}
27	7.542×10^{-3}	4.008×10^{-3}

It can be observed that these results are closer to those of the Bayes approach than those of the pseudo-Bayes approach. The Bayes approach gives as switching points $n = 18$ and 19, and the pseudo-Bayes approach gives switching points at 22–23 and 24–25, whereas here we found them at 21–22 and 23–24.

The considered data show that the recognition of a true state $J(t)$ using the last two considered approaches is very poor for high values

Table 8.5. Recognition criterion $\tilde{\Delta}_j$ for $J(0) = 2$ and $t = 10$.

n	$\tilde{\Delta}_1$	$\tilde{\Delta}_2$
10	0.025	0.033
11	0.018	0.025
12	0.012	0.019
13	6.674×10^{-3}	0.014
14	3.673×10^{-3}	8.900×10^{-3}
15	3.725×10^{-3}	5.185×10^{-3}
16	6.961×10^{-3}	2.790×10^{-3}
17	0.010	2.780×10^{-3}
18	0.013	5.382×10^{-3}
19	0.016	8.014×10^{-3}

of time t. Reducing time t sufficiently improves the results. This can be seen in Table 8.5, which is analogous to Table 8.4.

8.4 Markov-modulated Poisson process

We now consider a more general case, in which the external random environment is described as a continuous-time Markov chain $J(t)$ with k states. As before, $\lambda_{i,j}$ are the transition intensities between states i and j, $\lambda = (\lambda_{i,j})_{k\times k}$ is the corresponding matrix, and the Poisson flow $X(t)$ has an intensity α_i if the ith state of chain $J(t)$ occurs.

Our aim is to recognize state $J(t)$ if the initial state $J(0) = i$ and the value $X(t)$ are known. To this end, we use formula (2.7) to obtain the averages of the Poisson process $X(t)$ jointly with the probabilities of state $J(t)$.

The conditional expectation $E_i(X(0,t)|J(t) = j)$ with the conditions $J(0) = i$ and $J(t) = j$ is calculated as follows:

$$E_i(X(t)|J(t) = j) = \frac{E_J(X(t)_{i,j}}{P\{J(t) = j|J(0) = i\}}, \tag{8.13}$$

where probability $P\{J(t) = j|J(0) = i\}$ is calculated through formula (1.9).

Acting as earlier (see Definition 8.1), we calculate the recognition criterion:

$$\Delta_j = |n - E_i(X(t)|J(t) = j)|, \quad j = 1, \ldots, k, \tag{8.14}$$

find the minimal value Δ_{j*}, and assume j^* as state $J(t)$.

Consider a numerical example with the following input data: $k = 3$,

$$\lambda = \begin{pmatrix} 0 & 0.1 & 0.2 \\ 0.2 & 0 & 0.3 \\ 0.3 & 0.4 & 0 \end{pmatrix}, \quad \alpha = \begin{pmatrix} 1.2 \\ 1 \\ 0.5 \end{pmatrix}.$$

Tables 8.6–8.8 present the resulting values Δ_{j*} and j^* for $t = 10$ and different input states $J(0) = 1, 2, 3$.

The results show that monitoring is a difficult problem whose successful solution depends on the initial data. The initial data considered corresponded to favorable conditions because the different states were well distinguished: the arrival intensities for different states were 1.2, 1, 0.5. Additionally, stationary probabilities of the state J are: 0.451 for the first state, 0.294 for the second state, and 0.255 for the third state.

Table 8.6. Resulting values Δ_{j*} and j^* for $J(0) = 1$ and $t = 10$.

n	5	6	7	8	9	10	11	12
Δ_{j*}	4.473	3.473	2.473	1.473	0.473	0.165	1.165	2.165
j^*	3	3	3	3	3	2	2	2
n	19	20	21	22	23	24	25	26
Δ_{j*}	9.165	10.165	11.65	12.165	13.165	12.865	11.185	10.165
j^*	2	2	2	2	2	1	1	1

Table 8.7. Resulting values Δ_{j*} and j^* for $J(0) = 2$ and $t = 10$.

N	5	6	7	8	9	10	11	12
Δ_{j*}	3.992	2.992	1.992	0.992	0.850	0.071	1.071	2.071
j^*	3	3	3	3	3	1	1	1

Table 8.8. Resulting values Δ_{j*} and j^* for $J(0) = 3$ and $t = 10$.

N	5	6	7	8	9	10	11	12
Δ_{j*}	3.910	2.910	1.910	0.910	0.090	0.519	1.519	2.519
$j*$	2	2	2	2	2	1	1	1

The transfer intensities from the states are respectively: $0.1 + 0.2 = 0.3$, $0.2+0.3 = 0.5$, $0.3+0.4 = 0.7$. Therefore, the stationary intensities of outputs from states are $0.451\times0.3 = 0.541$, $0.294\times0.5 = 0.294$, and $0.255\times0.7 = 0.127$. However, the obtained results cannot be considered excellent.

8.5 Monitoring of a degenerating system

A degradation system is described as a continuous-time Markov chain $J(t)$ with a finite working state set $E = \{1, \dots, k\}$ and one-step transitions from state $i \in \{1, \dots, k-1\}$ to the state $i+1$. The intensity of such a transition is known and denoted as λ_i. A transition in state k gives an absorbing state.

The current state of the chain is not observed, and only an initial state $J(0) = 1$ is known. We want to forecast the state $J(t)$ at instant t. For this, we have complete information on the incoming Poisson claims flow $X(t)$ with different intensities. Namely, the flow intensity is equal to α_i if the chain $J(t)$ has the state i. Values $a_1 < a_2 <, \dots, < a_k$ are known to us.

We want to recognize state $J(t)$ at instant t for a known number of arrivals n within interval $(0, t)$.

First, we derive the state-probability equations. These formulas are simpler than the general formula (8.7), which uses the eigenvalues and eigenvectors. For the first initial state, we have the following probability $\mathrm{Pr}St(j, t)$ for state j at time t:

$$\begin{aligned}\mathrm{Pr}St(j,t) &= \exp(-t\lambda_1), \quad \text{if } j = 1,\\ &= \int_0^t \left(\prod_{i=1}^{j-1} \lambda_i\right) \sum_{i=1}^{j-1} \exp(-\lambda_i \tau) \left(\prod_{j=1, j\neq i}^{j-1} (\lambda_j - \lambda_i)\right)^{-1}\\ &\quad \times \exp(-\lambda_j(t-\tau))d\tau, \quad \text{if } j > 1. \end{aligned} \tag{8.15}$$

We want to refine this state probability by using additional information on the number of arrivals. Let $P_{j,n}(t)$ be the probability that the state of $J(t)$ is j and n arrivals occur till time $t \geq 0$. For the initial state, we have

$$P_{1,n}(t) = (\alpha_1 t)^n \frac{1}{n!} \exp(-t(\alpha_1 + \lambda_1)), \quad n = 0, 1, \ldots \tag{8.16}$$

Furthermore, the following equation applies for the other states $j > 1$ and $n = 0, 1, \ldots$:

$$P_{j,n}(t) = \int_0^t \sum_{m=0}^{n} P_{j-1,m}(\tau)\lambda_{j-1} \frac{(\alpha_j(t-\tau))^{\mathrm{n-m}}}{(n-m)!} \times \exp\left(-(t-\tau)(\alpha_j + \lambda_{\mathrm{j}})\right)d\tau. \tag{8.17}$$

These probabilities are sequentially calculated, allowing us to use the Bayes approach. The posterior probability of state $J(t) = j$ under the condition $X(t) = n$ is calculated as follows:

$$P\{J(t) = j | X(t) = n\} = \frac{P\{J(t) = j, X(t) = n\}}{P\{X(t) = n\}} = \frac{P_{j,n}(t)}{\sum_{i=1}^{k} P_{i,n}(t)}, \quad j = 1, \ldots, k. \tag{8.18}$$

We conclude this Section with a numerical example. Consider system $J(t)$ with four working states, $k = 4$. The intensities of the transitions between the states and Poisson flow arrivals are expressed as follows:

$$\lambda = \begin{pmatrix} 0.6 \\ 0.7 \\ 0.8 \\ 0.9 \end{pmatrix}, \quad \alpha = \begin{pmatrix} 1.0 \\ 1.5 \\ 2.0 \\ 2.5 \end{pmatrix}.$$

Figure 8.7 shows the probabilities $\mathrm{Pr}St(j,t)$ for different working states j of the system at time t. Table 8.9 lists the corresponding numerical values. The last row contains the probability that the system is working.

Table 8.10 shows the maximal probabilities of the states $\mathrm{MPr}St(t)$ and the corresponding numbers of these states j^* at different times t.

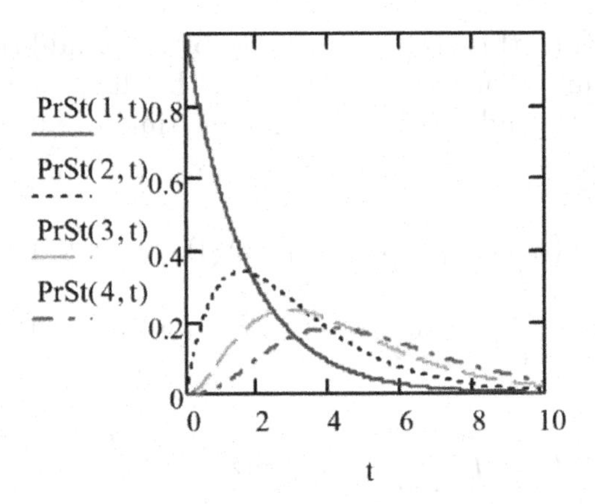

Fig. 8.7. Probabilities $\Pr St(j,t)$.

Table 8.9. Probabilities $\Pr St(j,t)$.

$j \setminus t$	0	1	2	3	4	5	6	7	8	9	10
1	1	0.549	0.301	0.165	0.091	0.050	0.027	0.015	0.008	0.005	0.002
2	0	0.313	0.328	0.257	0.179	0.118	0.074	0.045	0.027	0.016	0.009
3	0	0.104	0.208	0.233	0.207	0.162	0.117	0.080	0.052	0.033	0.021
4	0	0.026	0.100	0.161	0.182	0.170	0.141	0.107	0.077	0.053	0.035
$\sum$	1	0.992	0.937	0.816	0.659	0.500	0.359	0.247	0.164	0.107	0.067

Table 8.10. Maximum probabilities $\mathrm{MPr}St(t)$ and corresponding states j^*.

t	0	0.25	0.50	0.75	1.00	1.25	1.50	1.75	2.00	2.25	2.50
MPr	1	0.861	0.741	0.638	0.549	0.472	0.407	0.350	0.328	0.313	0.296
$j*$	1	1	1	1	1	1	1	1	2	2	2
t	2.75	3.00	3.25	3.50	3.75	4.00	4.25	4.50	4.75	5.00	5.25
MPr	0.277	0.257	0.237	0.224	0.216	0.207	0.197	0.185	0.175	0.170	0.164
$j*$	2	2	2	3	3	3	3	3	4	4	4

Now, we show how these probabilities can be refined using additional information on the number of arrivals $X(t)$. These probabilities can be calculated using formulas (8.16)–(8.18).

Table 8.11. Probabilities $P_{1,n}(t)$.

$n \setminus t$	0	0.25	0.50	0.75	1.00	1.25	1.50	1.75	2.00
0	1	0.670	0.449	0.301	0.202	0.135	0.091	0.061	0.041
1	0	0.168	0.225	0.226	0.202	0.169	0.136	0.106	0.082
2	0	0.020	0.056	0.05	0.101	0.106	0.102	0.093	0.082
3	0	0.002	0.009	0.021	0.034	0.044	0.051	0/054	0.054
4	0	0.000	0.001	0.004	0.08	0.014	0.019	0.024	0.027
5	0	0.000	0.000	0.000	0.002	0.003	0.006	0.008	0.011
6	0	0.000	0.000	0.000	0.000	0.000	0.001	0.002	0.004
$\sum$	1	0.861	0.741	0.638	0.549	0.472	0.407	0.350	0.301
$n \setminus t$	2.20	2.25	2.50	2.75	3.00	3.25	3.50	3.75	4.00
0	0.027	0.027	0.018	0.012	0.008	0.006	0.004	0.002	0.002
1	0.061	0.061	0.046	0.034	0.025	0.018	0.013	0.009	0.007
2	0.069	0.069	0.057	0.046	0.037	0.029	0.023	0.017	0.013
3	0.052	0.052	0.048	0.043	0.037	0.032	0.026	0.022	0.018
4	0.029	0.029	0.030	0.029	0.028	0.026	0.023	0.020	0.018
5	0.013	0.013	0.015	0.016	0.017	0.017	0.016	0.015	0.014
6	0.005	0.005	0.006	0.007	0.008	0.0099	0.009	0.010	0.009
$\sum$	0.259	0.259	0.223	0.192	0.165	0.141	0.121	0.104	0.081

Firstly, we provide an example of the presentation of the probabilities $P_{j,n}(t)$, calculated according to formula (8.16).

We produce the probabilities $P_{j,n}(t)$ calculated according to formula (8.16). These probabilities are shown in Table 8.11 for $j = 1$. The last row contains the probability sums of the corresponding columns.

Note that the sums of the probabilities for $t = 1, 2$, and 3 are 0.549, 0.301, and 0.165, respectively. They coincide with the probabilities of the states $\Pr St(1,t)$ from Table 8.9. This shows that the calculation result is correct. The sum for $t = 4$ is less than $\Pr St(1,4) = 0.091$, because not all significant values are presented in the column.

Let us fix an instant t and reduce the corresponding main results of the probabilities $P_{j,n}(t)$ in the table. The rows of this table correspond to the states of system j, whereas the columns correspond to the fixed number of arrivals n. Table 8.12 presents analogous results for $t = 3$. The maximum probabilities are highlighted in bold. The adopted value j for $J(t)$ corresponds to this probability.

Table 8.12. Probabilities $P_{j,n}(3)$.

$j \setminus n$	0	1	2	3	4	5	6
1	$\mathbf{8.23 \times 10^{-3}}$	**0.025**	0.037	0.037	0.028	0.017	8.33×10^{-3}
2	6.87×10^{-3}	0.024	**0.044**	**0.053**	**0.048**	0.036	0.023
3	3.10×10^{-3}	0.013	0.027	0.039	0.042	**0.038**	**0.028**
4	1.22×10^{-3}	5.65×10^{-3}	0.013	0.021	0.026	0.027	0.023

$j \setminus n$	7	8	9	10	11	12	13
1	3.57×10^{-3}	1.34×10^{-3}	4.46×10^{-4}	1.34×10^{-4}	3.65×10^{-5}	9.13×10^{-6}	2.11×10^{-6}
2	0.012	6.00×10^{-3}	2.61×10^{-3}	1.03×10^{-3}	3.75×10^{-4}	1.26×10^{-4}	3.95×10^{-5}
3	**0.019**	**0.011**	5.30×10^{-3}	2.54×10^{-3}	1.12×10^{-3}	4.62×10^{-4}	1.77×10^{-4}
4	0.017	**0.011**	$\mathbf{6.83 \times 10^{-3}}$	$\mathbf{3.76 \times 10^{-3}}$	$\mathbf{1.91 \times 10^{-3}}$	$\mathbf{9.04 \times 10^{-4}}$	$\mathbf{4.00 \times 10^{-4}}$

The sum of the maximum probabilities gives the probability that the recognition of the true state is correct. In our case, this probability is as follows:

$$\begin{aligned}8.23 \times 10^{-3} &+ 0.025 + 0.044 + 0.053 + 0.048 + 0.038 + 0.028\\ &+ 0.019 + 0.011 + 6.83 \times 10^{-3} + 3.76 \times 10^{-3}\\ &+ 1.91 \times 10^{-3} + 9.04 \times 10^{-4} + 4.00 \times 10^{-4} = 0.289.\end{aligned}$$

We see from Table 8.10 that for time $t = 3$, the second state $j = 2$ is the accepted solution if additional information is missing. This yielded a probability of 0.257, which was true for this solution. Our results increased this probability to 0.289.

References

[1] Cham. Yu, J., Qin, S.J., Multimode process monitoring with Bayesian inference-based finite Gaussian mixture models. *AIChE Journal*, 2008;54(7):1811–1829.

[2] Cuentas, S., Peñabaena-Niebles, R., Garcia, E., Support vector machine in statistical process monitoring: A methodological and analytical review. *The International Journal of Advanced Manufacturing Technology*, 2017;91(1-4):485–500.

[3] Yu, W., Zhao, C., Recursive exponential slow feature analysis for fine-scale adaptive processes monitoring with comprehensive operation status identification. *IEEE Transactions on Industrial Informatics*, 2018;15(6):3311–3323.

[4] Bartolucci, F., Farcomeni, A., Pennoni, F., *Latent Markov Models for Longitudinal Data*. Chapman & Hall/CRC, Boca Raton, 2013.

[5] Fischer, W., Meier-Hellstern, K., The Markov modulated Poisson process cookbook. *Performance Evaluation*, 1992;18:149–171.

[6] Kijima, M., *Markov Processes for Stochastic Modeling*. The University Press, Cambridge, UK, 1997.

[7] Mamon, R.S., R.J. Elliot (eds.), *Hidden Markov Models in Finance*, International Series in Operations Research & Management Science, Vol. 104. Springer, New York, 2007.

[8] Özekici, S., Soyer, R., Semi-Markov modulated Poisson process: Probabilistic and statistical analysis. *Mathematical Methods Operational Research*, 2006;64:125–144.

[9] Pacheco, A., Tang, L.C., Prabhu, N.U., *Markov-Modulated Processes & Semiregenerative Phenomena*. World Scientific, New Jersey, London, 2009.

[10] Petropoulos, A., Chatzis, S.P., Xanthopoulos, S., A novel corporate credit rating system based on Student's-t hidden Markov models. *Expert Systems with Applications*, 2016;53:87–105. doi:10.1016/j.eswa.2016.01.015.

Chapter 9

Reliability System *k-out-of-n* in an External Random Environment

9.1 Introduction

The *k-out-of-n model* is a popular reliability model (see list References). It is assumed that the considered system consists of n parallel components and fails when at least k of its components fail, which can be repaired using a single or multiple servers. Various scenarios for repairing entire systems have been considered [1,2,16,17]. Full repair of the system implies that the repaired system functions as a new one.

It was often assumed that the lifetimes and repair times of the elements had exponential distributions. The full repair time can have a general distribution [1,3–6].

Generally, the considered random variables are supposed to be mutually independent; however, their dependence is also considered [7]. Non-identical components have been considered in previous studies [8,9,15].

We considered such a system in an external random environment represented by a finite continuous-time irreducible Markov chain $J(t)$, $t \geq 0$ (see Chapter 1), which is defined by intensities of transitions from the ith to jth state $\lambda_{i,j} \geq 0$, $i, j = 0, \ldots, m-1$. Supposedly, the sojourn time in state i has an exponential distribution with the parameter $\Lambda_i = \sum_{j=0}^{m-1} \lambda_{i,j}$ and when it ends, the chain transfers to the state j with probability $\lambda_{i,j}/\Lambda_i$.

If the Markov chain is in the ith state, where $i = 0, \dots, m-1$, then the lifetimes of the components are independent random variables, having exponential distribution with identical intensity α_i.

The following repair variants were considered. *The first variant* assumes the absence of repairs (see Section 9.2); thus, it is necessary to determine the time distribution until failure. *The second variant* assumes (see Section 9.4) that failed components are repaired by s servers. If Markov chain is in the ith state, $i = 0, \dots, m-1$, then a repair time has the exponential distribution with intensity β_i independent of other random variables. The first-in-first-out (FIFO) service discipline was used. *The third variant* supposes (see Section 9.5) that the entire system's capital repair is performed after its failure, and the duration of capital repairs had a general distribution.

Numerical examples illustrate analytical results (see Sections 9.3 and 9.6).

9.2 Reliability without a repair

We consider the two-dimensional random process $Z(t) = (J(t), X(t))$, where $J(t) \in \{0, \dots, m-1\}$ is a state of Markov chain, and $X(t) \in \{0, \dots, k-1\}$ is the number of failed components at instant $t \geq 0$, $X(0) = x$. We wish to calculate probabilities $P_{(i,x),(j,y)} = P\{Z(t) = (j, y)|J(0) = i, X(0) = x\}$.

Considering the first jump after the initial time moment 0, we deduce that a small time interval $\Delta t > 0$ leads to the following equation system:

$$\begin{aligned}
&P_{(i,x),(j,y)}(t + \Delta t)\\
&\quad = (1 - \Delta t(\Lambda_i + (n-x)\alpha_{\mathrm{i}}))P_{(i,x),(j,y)}(t) + \sum_{\nu=1}^{k} \lambda_{i,v}\Delta P_{(\nu,x),(j,y)}(t)\\
&\qquad + \Delta t(n-x)\alpha_{\mathrm{i}} P_{(i,x+1),(i,y)}(t) + o(\Delta t), \quad t \geq 0; \;\; \forall i, j;\\
&\qquad x \in \{0, \dots, k-1\}; \;\; y \in \{x, \dots, k-1\},
\end{aligned}$$

where $P_{(i,k),(i,k)} = 0$.

Hence, we obtain the following system of differential equations with constant coefficients:

$$\begin{aligned}\dot{P}_{(i,x),(j,y)}(t) &= -(\Lambda_i + (n-x)\alpha_{\mathrm{i}})P_{(i,x),(j,y)}(t) \\ &\quad + \sum_{\nu=1}^{k} \lambda_{i,v} P_{(\nu,x),(j,y)}(t) + (n-x)\alpha_{\mathrm{i}} P_{(i,x+1),(i,y)}(t), \\ &\quad t \geq 0, \ \forall i,j; \quad x \in \{0,\dots,k-2\}; \\ &\quad y \in \{x,\dots,k-2\}, \\ \dot{P}_{(i,k-1),(j,k-1)}(t) &= -(\Lambda_i + (n-(k-1))\alpha_{\mathrm{i}})P_{(i,k-1),(j,k-1)}(t) \\ &\quad + \sum_{\nu=1}^{k} \lambda_{i,v} P_{(\nu,k-1),(j,k-1)}(t), \quad t \geq 0, \ \forall i,j. \end{aligned} \tag{9.1}$$

These equations provide the following expressions for the nonzero elements of generator G:

$$\begin{aligned} G_{(i,x),(j,x)} &= \lambda_{i,j}, \quad j \neq i; \\ G_{(i,x),(i,x+1)} &= \alpha_i(n-x), \quad x = 0,\dots,k-2; \\ G_{(i,x),(i,x)} &= -(\Lambda_i + \alpha_i(n-x)), \quad x = 0,\dots,k-1. \end{aligned} \tag{9.2}$$

To represent this system in a matrix form, it is necessary to enumerate the states of the two-dimensional process $Z(t)$. The total number of states is denoted by *mk*. It is convenient to represent all states by matrix of dimension $m \times k$. Rows of the matrix contain states (i, x), with the same value of i. Columns contain the states with the same value of x. Because the states are enumerated through columns, the state (i, x), $i \in \{0,\dots,m-1\}$, $x \in \{0,\dots,k-1\}$, has the following number which uniquely determines the state:

$$Num(i,x) = mx + i, \quad i \in \{0,\dots,m-1\}, \quad x \in \{0,\dots,k-1\}. \tag{9.3}$$

Let $\mathrm{mod}(x, y)$ return the remainder on dividing x by y. Then, the following states (i, x) correspond to *Num*:

$$i = \mathrm{mod}(Num, m), \tag{9.4}$$

$$x = \frac{(Num - i)}{m}. \tag{9.5}$$

Furthermore, we suppose that the generator G of dimension $mk \times mk$ has the described enumeration of its own states. We use its eigenvalues $\chi_0, \chi_2, \ldots, \chi_{km-1}$ and eigenvectors $\beta_0, \beta_2, \ldots, \beta_{km-1}$, matrix of eigenvectors $B = (\beta_0, \ldots, \beta_{km-1})$, and its inverse matrix $B^{-1} = \tilde{B} = (\tilde{\beta}_0^T, \ldots, \tilde{\beta}_{km-1}^T)^T$ with rows $\tilde{\beta}_0, \ldots, \tilde{\beta}_{km-1}$. Herein, all values $\chi_0, \chi_2, \ldots, \chi_{km-1}$ are different and negative.

Now, the matrix

$$P(t) = \sum_{\varsigma=0}^{km-1} \exp(\chi_\varsigma t)\beta_\varsigma \tilde{\beta}_\varsigma, \quad t \geq 0, \tag{9.6}$$

gives the probability that failure is absent until time t for different initial and final states.

The probability that failure is absent until time t if the initial state i occurs is calculated as follows:

$$\mathrm{Pr}_i(t) = \sum_j P_{i,j}(t), \quad t \geq 0, \quad i = 0, \ldots, m-1. \tag{9.7}$$

It is interesting to compare this result with the case in which the system's elements fail independently. The necessary results for a single element can be obtained from Chapter 7. Now, we interpret "the arrival" as a failure of the considered element. Instead of the vector μ of arrival's intensities, the vector $\alpha = (\alpha_0 \ \cdots \ \alpha_{m-1})$ of the failure's intensities is used. Equation (7.15) gives the probability $F_i(1,t)$ that the element fails until time t, if the ith state initially occurs. Therefore, the probability that our system *k-out-of-n* operates at instant t is calculated as follows:

$$R_i(t) = \sum_{x=0}^{k-1} \frac{n!}{x!(n-x)!} F_i(1,t)^x (1 - F_i(1,t))^{n-x}, \quad t \geq 0. \tag{9.8}$$

The following example illustrates the presented results.

9.3 Example

Similar to the example in the previous chapter, it is assumed that the number of Markov chain states $m = 3$ and the matrix of intensities of transitions is as follows:

$$\lambda = \begin{pmatrix} 0 & 0.4 & 1.0 \\ 0.7 & 0 & 1.1 \\ 0 & 1.5 & 0 \end{pmatrix}.$$

Intensities of failures for different states are determined by vector

$$\alpha = (0.5 \quad 0.75 \quad 1)^T.$$

The system consists of $n = 7$ components and fails when at least $k = 4$ of its components fail. The number of system states was $km = 12$. Table 9.1 contains the numbers of these states.

Generator G of system (9.2) is shown in Fig. 9.1.

Table 9.1. Enumeration of states.

	$x = 0$	$x = 1$	$x = 2$	$x = 3$
$i = 0$	0	3	6	9
$i = 1$	1	4	7	10
$i = 2$	2	5	8	11

$$G := \begin{pmatrix}
-4.9 & 0.4 & 1 & 3.5 & 0 & 0 & 0 & 0 & 0 & 0 & 0 & 0 \\
0.7 & -7.05 & 1.1 & 0 & 5.25 & 0 & 0 & 0 & 0 & 0 & 0 & 0 \\
0 & 1.5 & -8.5 & 0 & 0 & 7 & 0 & 0 & 0 & 0 & 0 & 0 \\
0 & 0 & 0 & -4.4 & 0.4 & 1 & 3 & 0 & 0 & 0 & 0 & 0 \\
0 & 0 & 0 & 0.7 & -6.3 & 1.1 & 0 & 4.5 & 0 & 0 & 0 & 0 \\
0 & 0 & 0 & 0 & 1.5 & -7.5 & 0 & 0 & 6 & 0 & 0 & 0 \\
0 & 0 & 0 & 0 & 0 & 0 & -3.9 & 0.4 & 1 & 2.5 & 0 & 0 \\
0 & 0 & 0 & 0 & 0 & 0 & 0.7 & -5.55 & 1.1 & 0 & 3.75 & 0 \\
0 & 0 & 0 & 0 & 0 & 0 & 0 & 1.5 & -6.5 & 0 & 0 & 5 \\
0 & 0 & 0 & 0 & 0 & 0 & 0 & 0 & 0 & -3.4 & 0.4 & 1 \\
0 & 0 & 0 & 0 & 0 & 0 & 0 & 0 & 0 & 0.7 & -4.8 & 1.1 \\
0 & 0 & 0 & 0 & 0 & 0 & 0 & 0 & 0 & 0 & 1.5 & -5.5
\end{pmatrix}$$

Fig. 9.1. The generator G.

The vector of eigenvalues is as follows:

$$\chi^T := (-4.624 \quad -6.645 \quad -9.18 \quad -4.066 \quad -5.896 \quad -8.238 \quad -3.486 \quad -5.161 \quad -7.303 \quad -2.875 \quad -4.449 \quad -6.376)$$

Matrix B of the eigenvalues and its inverse matrix B^{-1} are presented in Figs. 9.2 and 9.3, respectively.

Now, we can calculate the probabilities $\Pr_i(t)$ that failure is absent until time t for different initial states $i = 0, 1,$ and 2 (see formula (9.7)). The corresponding graphs are shown in Fig. 9.4.

We conclude the considered example by comparing these results with a case in which the system's elements fail independently (see formula (9.8)). Figure 9.5 shows the full uninspected concurrence

$$B := \begin{pmatrix} -0.936 & 0.474 & -0.172 & -0.908 & 0.541 & -0.183 & -0.854 & 0.601 & -0.192 & -0.774 & 0.644 & -0.198 \\ -0.328 & -0.685 & -0.407 & -0.362 & -0.606 & -0.43 & -0.394 & -0.499 & -0.444 & -0.418 & -0.37 & -0.448 \\ -0.127 & -0.554 & 0.897 & -0.158 & -0.566 & 0.873 & -0.196 & -0.559 & 0.83 & -0.239 & -0.528 & 0.768 \\ 0 & 0 & 0 & -0.13 & 0.077 & -0.026 & -0.244 & 0.172 & -0.055 & -0.332 & 0.276 & -0.085 \\ 0 & 0 & 0 & -0.052 & -0.087 & -0.061 & -0.113 & -0.143 & -0.127 & -0.179 & -0.158 & -0.192 \\ 0 & 0 & 0 & -0.023 & -0.081 & 0.125 & -0.056 & -0.16 & 0.237 & -0.102 & -0.226 & 0.329 \\ 0 & 0 & 0 & 0 & 0 & 0 & -0.041 & 0.029 & -0.009 & -0.111 & 0.092 & -0.028 \\ 0 & 0 & 0 & 0 & 0 & 0 & -0.019 & -0.024 & -0.021 & -0.06 & -0.053 & -0.064 \\ 0 & 0 & 0 & 0 & 0 & 0 & -0.009 & -0.027 & 0.04 & -0.034 & -0.075 & 0.11 \\ 0 & 0 & 0 & 0 & 0 & 0 & 0 & 0 & 0 & -0.022 & 0.018 & -0.006 \\ 0 & 0 & 0 & 0 & 0 & 0 & 0 & 0 & 0 & -0.012 & -0.011 & -0.013 \\ 0 & 0 & 0 & 0 & 0 & 0 & 0 & 0 & 0 & -0.007 & -0.015 & 0.022 \end{pmatrix}$$

Fig. 9.2. Matrix of eigenvalues B.

$$BIn := \begin{pmatrix} -0.899 & -0.354 & -0.332 & 6.294 & 2.477 & 2.327 & -18.883 & -7.432 & -6.982 & 31.472 & 12.386 & 11.636 \\ 0.37 & -0.923 & -0.348 & -2.592 & 6.461 & 2.435 & 7.775 & -19.384 & -7.304 & -12.958 & 32.306 & 12.173 \\ 0.101 & -0.62 & 0.853 & -0.709 & 4.337 & -5.971 & 2.128 & -13.012 & 17.912 & -3.547 & 21.686 & -29.853 \\ 0 & 0 & 0 & -6.083 & -2.903 & -2.701 & 36.498 & 17.417 & 16.207 & -91.246 & -43.543 & -40.519 \\ 0 & 0 & 0 & 3.027 & -6.47 & -2.55 & -18.164 & 38.821 & 15.299 & 45.411 & -97.052 & -38.247 \\ 0 & 0 & 0 & 0.862 & -4.724 & 5.874 & -5.17 & 28.347 & -35.245 & 12.925 & -70.866 & 88.113 \\ 0 & 0 & 0 & 0 & 0 & 0 & -17.583 & -10.405 & -9.631 & 87.916 & 52.027 & 48.153 \\ 0 & 0 & 0 & 0 & 0 & 0 & 10.995 & -19.812 & -8.066 & -54.976 & 99.059 & 40.33 \\ 0 & 0 & 0 & 0 & 0 & 0 & 3.25 & -15.802 & 17.599 & -16.252 & 79.008 & -87.997 \\ 0 & 0 & 0 & 0 & 0 & 0 & 0 & 0 & 0 & -27.94 & -20.943 & -19.422 \\ 0 & 0 & 0 & 0 & 0 & 0 & 0 & 0 & 0 & 22.968 & -34.423 & -14.177 \\ 0 & 0 & 0 & 0 & 0 & 0 & 0 & 0 & 0 & 7.096 & -30.165 & 29.793 \end{pmatrix}$$

Fig. 9.3. Inverse matrix B^{-1}.

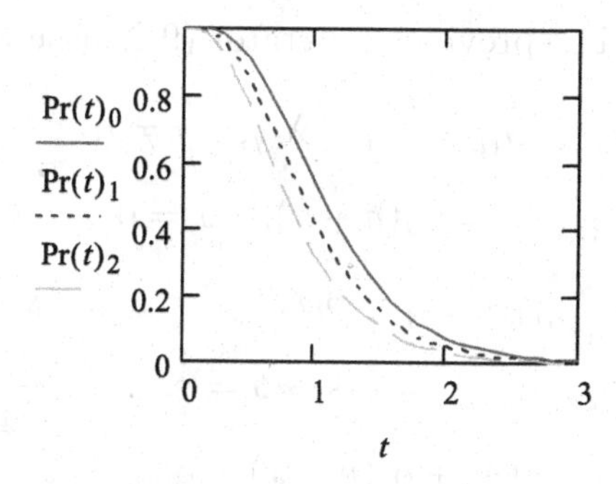

Fig. 9.4. Probabilities of the worked system at time t.

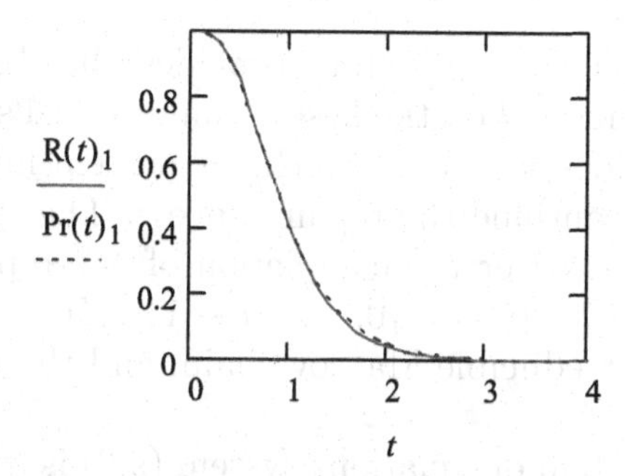

Fig. 9.5. Probabilities of the worked system at time t.

of the results for $i = 1$, which confirms the correctness of the used approach and renders it applicable in more complex situations.

9.4 Reliability with repairs

Now, we suppose that $k > 1$ and s is the number of servers, such that $s = 1, \ldots, k-1$. The FIFO service discipline was used. If the random environment is in the ith state, the servers' repair times are independent and exponentially distributed random variables with intensity β_i. We wish to determine the time distribution until system failure.

As earlier, the two-dimension random process $Z(t) = (J(t), X(t))$ is conceded, where $J(t) \in \{0, \ldots, m-1\}$ is the state of Markov chain, $X(t) \in \{0, \ldots, k-1\}$ is the number of failed components at instant

$t \geq 0$, $X(0) = x$. The previous generator (9.2) assumes the following:

$$G_{(i,x),(j,x)} = \lambda_{i,j}, \quad j \neq i;$$

$$G_{(i,x),(i,x+1)} = \alpha_i(n - x), \quad x = 0, \ldots, k-2;$$

$$G_{(i,x),(i,x-1)} = \beta_i x, \quad x = 1, \ldots, s;$$

$$G_{(i,x),(i,x-1)} = \beta_i s, \quad x = \mathrm{s}, \ldots, k-1;$$

$$G_{(i,x),(i,x)} = -(\Lambda_i + \alpha_i(n - x) + \beta_i x), \quad x = 1, \ldots, s.$$

$$G_{(i,x),(i,x)} = -(\Lambda_i + \alpha_i(n - x) + \beta_i s), \quad x = s, \ldots, k-1. \tag{9.9}$$

The aforementioned procedure provides the distributions (9.6) and (9.7) of the time when the system failure is absent.

Now, we consider a case in which the system fails in state (i, t), but all s servers continue the repair process. The system renews its capacity to operate after the completion of one repair. The process $Z(t) = (J(t), X(t))$, $J(t) \in \{0, \ldots, m-1\}$, $X(t) \in \{0, \ldots, k\}$ is a continuous-time irreducible Markov chain, and the number of states is $m(k+1)$.

Herein, the system of equations system (9.9) is replenished by the following non-zero elements for $i = 0, \ldots, m-1$:

$$\begin{aligned} G_{(i,k-1),(i,k)} &= \alpha_i(n - (k-1)), \\ G_{(i,k),(i,k-1)} &= \beta_i s, \\ G_{(i,k),(i,k)} &= -\beta_i s. \end{aligned} \tag{9.10}$$

Stationary distribution of states probabilities is calculated as it is described in Chapter 1.

9.5 Capital repair of the system

We now consider the case in which capital repair of the entire system is performed after its failure. Let $h_i(t)$ be the density of the capital repair duration if the random environment is initially in the ith state. It is assumed that the components do not fail during capital repair, and all n elements of the system are available after capital repair.

We consider the stationary distribution of the process $\tilde{Z}(t) = (J(t), \tilde{X}(t))$, where $J(t) \in \{0, \dots, m-1\}$ is a state of Markov chain, $\tilde{X}(t) \in \{0, \dots, k\}$ is the number of the failed components at the instant $t \geq 0$, $X(0) = 0$. The state $\tilde{X}(t) = k$ indicates that the system has capital repairs.

Notably, the process $\tilde{Z}(t)$ is not Markovian. We wish to calculate the stationary probabilities of these process states as follows:

$$ps_{i,x} = \lim_{t\to\infty} P\{\tilde{Z}(t) = (i,x)\}, \quad i = 1, \dots, m-1; \; x = 0, \dots, k.$$

Let us consider the embedded Markov chain that is built on the time's moments of jumps of the process $\tilde{Z}(t)$, $t \geq 0$. We denote $Q_{(i,x),(j,y)}$ the transfer probability per one step from the state (i,x) to the state (j,y).

Probabilities of random environment transition from the state i to the state j for constant value x are calculated as follows: for i, $j = 0, \dots, m-1$,

$$Q_{(i,x),(j,x)} = \frac{\lambda_{i,j}}{\Lambda_i + (n-x)\alpha_i + \beta_i \min(x,s)}, \quad x = 0, \dots, k-1. \tag{9.11}$$

Furthermore,

$$Q_{(i,x),(i,x+1)} = \frac{(n-x)\alpha_i}{\Lambda_i + (n-x)\alpha_i + \beta_i \min(x,s)}, \quad x = 0, \dots, k-1, \tag{9.12}$$

$$Q_{(i,x),(i,x-1)} = \frac{\beta_i \min(x,s)}{\Lambda_i + (n-x)\alpha_i + \beta_i \min(x,s)}, \quad x = 0, \dots, k-1, \tag{9.13}$$

The probabilities $\tilde{P}_{i,j}(t)$, $j = 1, \dots, m$ of a random environment transition from state i to state j during the capital repair time are given by formula (1.9). Therefore, the probability that capital repair ends in state j is given as follows:

$$Q_{(i,k),(j,0)} = \int_0^\infty \tilde{P}_{i,j}(t) h_i(t) dt. \tag{9.14}$$

The other probabilities $Q_{(i,x),(j,y)}$ equal zero.

The stationary probabilities $q = (q_{0,0}, \ldots, q_{m-1,k})$ for the matrix Q are founded as usual. Because

$$q = qQ,$$

q^T is an eigenvector of matrix Q^T, corresponding to its unit eigenvalue. Operation of the normalization is necessary because an eigenvector is determined up to a constant factor.

The process $\tilde{Z}(t) = (J(t), \tilde{Z}(t))$, where $J(t) \in \{0, \ldots, m-1\}$, and $\tilde{Z}(t) \in \{0, \ldots, k\}$ is a continuous time semi-Markov process [4,10–12,19]. To calculate the stationary probabilities of the states, we use stationary probabilities q and the mean sojourn times in various states. We denote $T(i, x)$ as the sojourn time in state (i, x) per one visit. For $i = 0, \ldots, m-1$, we have the following:

$$E(T(i,k)) = \int_0^\infty t h_i(t) dt, \tag{9.15}$$

$$E(T(i,x)) = \frac{1}{\Lambda_i + (n-x)\alpha_i + \beta_i \min(x,s)} x = 0, \ldots, k-1. \tag{9.16}$$

A sojourn time TW in the set of worked states has the following mean value:

$$E(TW) = \sum_{i=0}^{m-1} \sum_{x=0}^{k-1} q_{i,x} E(T(i,x)). \tag{9.17}$$

The mean sojourn time in the set of failed states is calculated as follows:

$$E(TF) = \sum_{i=0}^{m-1} q_{i,k} E(T(i,k)). \tag{9.18}$$

Finally, we have the following stationary probabilities of states:

$$ps_{i,x} = \frac{q_{i,x} E(T(i,x))}{E(TW) + E(TF)}, \quad i = 1, \ldots, m-1; \; x = 0, \ldots, k-1. \tag{9.19}$$

The stationary probability that the system is worked is calculated as follows:

$$R = \frac{E(TW)}{E(TW) + E(TF)}. \tag{9.20}$$

In conclusion, we conducted a simple profit analysis that is prevalent in the literature [13,18]. Let the expense on one server be c euro per unit time, and let the income per unit time of the worked system be r euro. We wish to obtain the number of servers s^* yielding the maximal mean profit in the stationary case.

Mean profit as a function of server number s is calculated as follows:

$$E(\mathrm{In}(s)) = rR(s) - cs, \tag{9.21}$$

where $R(s)$ is calculated by formula (9.20).

For a positive profit, it is necessary that $s \leq rR(s)/c$. The optimal value of s is determined through an item-by-item examination.

9.6 Example (continuation)

Let the number of servers be equal to 2 ($s = 2$) and the vector of repair intensities by one server for different states of the random environment be as follows:

$$\beta = (1 \quad 2 \quad 3)^T.$$

The generator (9.9), matrix B of the eigenvectors, and inverse matrix B^{-1} are presented in Figs. 9.6–9.8.

$$\mathrm{Gen} := \begin{pmatrix} -4.9 & 0.4 & 1 & 3.5 & 0 & 0 & 0 & 0 & 0 & 0 & 0 & 0 \\ 0.7 & -7.05 & 1.1 & 0 & 5.25 & 0 & 0 & 0 & 0 & 0 & 0 & 0 \\ 0 & 1.5 & -8.5 & 0 & 0 & 7 & 0 & 0 & 0 & 0 & 0 & 0 \\ 1 & 0 & 0 & -5.4 & 0.4 & 1 & 3 & 0 & 0 & 0 & 0 & 0 \\ 0 & 2 & 0 & 0.7 & -8.3 & 1.1 & 0 & 4.5 & 0 & 0 & 0 & 0 \\ 0 & 0 & 3 & 0 & 1.5 & -10.5 & 0 & 0 & 6 & 0 & 0 & 0 \\ 0 & 0 & 0 & 2 & 0 & 0 & -5.9 & 0.4 & 1 & 2.5 & 0 & 0 \\ 0 & 0 & 0 & 0 & 4 & 0 & 0.7 & -9.55 & 1.1 & 0 & 3.75 & 0 \\ 0 & 0 & 0 & 0 & 0 & 6 & 0 & 1.5 & -12.5 & 0 & 0 & 5 \\ 0 & 0 & 0 & 0 & 0 & 0 & 2 & 0 & 0 & -5.4 & 0.4 & 1 \\ 0 & 0 & 0 & 0 & 0 & 0 & 0 & 4 & 0 & 0.7 & -8.8 & 1.1 \\ 0 & 0 & 0 & 0 & 0 & 0 & 0 & 0 & 6 & 0 & 1.5 & -11.5 \end{pmatrix}$$

Fig. 9.6. The generator G.

$$B := \begin{pmatrix}
0.026 & 0.037 & -0.084 & 0.387 & -0.139 & 0.429 & -0.113 & -0.872 & -0.576 & -0.571 & -0.835 & 0.355 \\
0.059 & -0.343 & -0.21 & 0.362 & 0.691 & 0.184 & -0.356 & 8.781\times 10^{-3} & 0.5 & -0.098 & -0.167 & 0.38 \\
-0.29 & -0.025 & 0.651 & 0.34 & 0.208 & 0.037 & 0.707 & 0.102 & 0.39 & 0.319 & -0.182 & -0.414 \\
-0.042 & -0.059 & 0.046 & 0.34 & 0.055 & -0.554 & -0.057 & 0.306 & -0.034 & -0.469 & -0.195 & 0.237 \\
-0.096 & 0.516 & 0.134 & 0.323 & -0.383 & -0.139 & -0.06 & 0.096 & 0.122 & -0.075 & 0.046 & 0.31 \\
0.49 & 0.097 & -0.424 & 0.307 & -0.186 & -0.043 & 0.115 & 0.031 & 0.048 & 0.294 & -0.087 & -0.389 \\
0.053 & 0.075 & 0.026 & 0.271 & 0.079 & 0.584 & 0.059 & 0.181 & 0.163 & -0.349 & 0.196 & 0.135 \\
0.124 & -0.621 & 0.033 & 0.266 & -0.145 & 0.042 & 0.137 & -0.015 & -0.158 & -0.052 & 0.172 & 0.234 \\
-0.659 & -0.188 & -0.144 & 0.256 & -0.031 & 6.531\times 10^{-3} & -0.293 & -0.053 & -0.187 & 0.25 & -0.018 & -0.336 \\
-0.034 & -0.049 & -0.064 & 0.16 & -0.131 & -0.327 & 0.088 & -0.246 & 0.14 & -0.189 & 0.302 & 0.048 \\
-0.082 & 0.382 & -0.141 & 0.165 & 0.411 & 0.041 & 0.191 & -0.134 & -0.268 & -0.026 & 0.184 & 0.132 \\
0.446 & 0.161 & 0.528 & 0.164 & 0.25 & 0.043 & -0.435 & -0.099 & -0.264 & 0.163 & 0.022 & -0.222
\end{pmatrix}$$

Fig. 9.7. Matrix B.

$$B^{-1} := \begin{pmatrix}
-2.16\times 10^{-3} & 0.036 & -0.134 & 8.705\times 10^{-3} & -0.144 & 0.531 & -0.012 & 0.197 & -0.715 & 7.219\times 10^{-3} & -0.114 & 0.404 \\
0.015 & -0.128 & -0.031 & -0.063 & 0.526 & 0.108 & 0.091 & -0.736 & -0.13 & -0.056 & 0.439 & 0.065 \\
0.015 & -0.161 & 0.405 & -0.021 & 0.214 & -0.626 & -0.016 & 0.167 & -0.191 & 0.034 & -0.335 & 0.631 \\
0.091 & 0.21 & 0.235 & 0.245 & 0.485 & 0.513 & 0.261 & 0.443 & 0.444 & 0.172 & 0.255 & 0.245 \\
-0.104 & 0.405 & 0.122 & 0.225 & -0.621 & -0.165 & -0.091 & -0.263 & -0.087 & -0.069 & 0.755 & 0.189 \\
0.133 & -0.016 & -0.013 & -0.554 & -1.201\times 10^{-4} & -0.036 & 0.808 & 0.09 & 0.099 & -0.518 & -0.109 & -0.06 \\
0.059 & -0.305 & 0.493 & 0.025 & -0.218 & 0.155 & -0.06 & 0.33 & -0.478 & -0.086 & 0.549 & -0.554 \\
-0.438 & -0.204 & -0.104 & 0.734 & 0.084 & 0.143 & 0.426 & 0.184 & 0.085 & -1.239 & -0.017 & -0.159 \\
-0.079 & 0.585 & 0.249 & -0.345 & 0.219 & 0.043 & 0.117 & -0.609 & -0.239 & 0.741 & -0.752 & -0.222 \\
-0.221 & 0.386 & 0.131 & -0.797 & 0.82 & 0.195 & -1.051 & 0.687 & 0.088 & -0.823 & 0.358 & 1.164\times 10^{-3} \\
-0.419 & -0.184 & -0.255 & -0.339 & -0.025 & -0.194 & 0.468 & 0.319 & 0.142 & 0.893 & 0.389 & 0.263 \\
-9.801\times 10^{-3} & 0.461 & -0.166 & -0.338 & 0.989 & -0.453 & -0.683 & 0.834 & -0.475 & -0.67 & 0.435 & -0.313
\end{pmatrix}$$

Fig. 9.8. Inverse matrix B^{-1}.

The following is the vector of eigenvalues:

$$\begin{aligned}\chi^T := (&-20.634 \quad -14.928 \quad -13.54 \quad -0.581 \quad -9.769 \quad -9.158 \\ &-8.121 \quad -6.249 \quad -5.716 \quad -2.514 \quad -3.784 \quad -3.306)\end{aligned}$$

Figure 9.9 contains the graph of the distributions (9.7) with the generator (9.9) for the time when the system failure is absent. A comparison with the corresponding graph in Fig. 9.4 indicates that these probabilities are sufficiently large.

Furthermore, we consider the case in which capital repairs occur. Let the duration of the system's capital repair have a Weibull distribution [14] with density.

$$h_i(t) = \alpha_i t^{\alpha_i - 1}\exp(-t^{\alpha_i}), \quad t \geq 0, \;\; i = 0, \ldots, 2,$$

where $\alpha_i = 1 + 0.5i$.

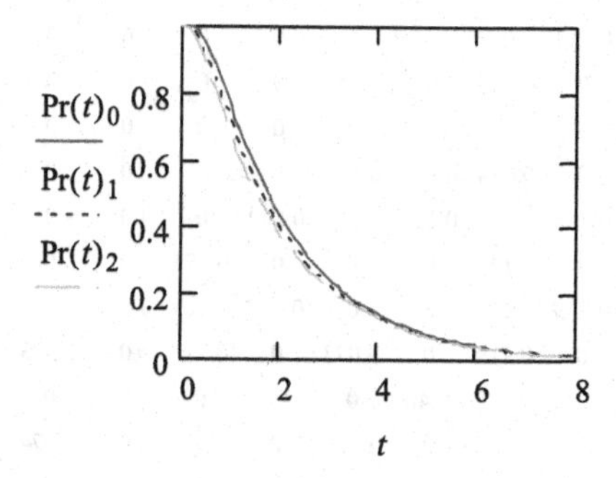

Fig. 9.9. Distributions (9.7) of time when the system failure is absent.

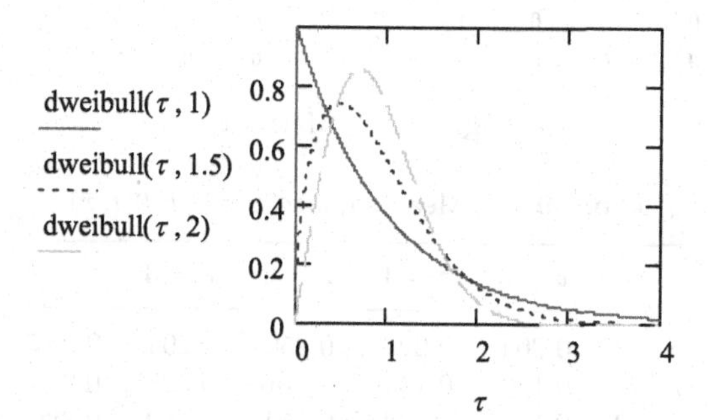

Fig. 9.10. Densities $h_i(t) = \text{dweibull}(t, \alpha_i)$.

Table 9.2. Enumeration of states.

	$x = 0$	$x = 1$	$x = 2$	$x = 3$	$x = 4$
$i = 0$	0	3	6	9	12
$i = 1$	1	4	7	10	13
$i = 2$	2	5	8	11	14

Figure 9.10 contains the corresponding graph, where $h_i(t) = dweibull(t, \alpha_i)$.

The number of states of the system is $(k+1)m = 15$. Table 9.2 contains the enumeration of the states.

Matrix Q of the transition probabilities between states of the embedded Markov chain is presented in Fig. 9.11. The stationary

$$Q := \begin{pmatrix}
0 & 0.082 & 0.204 & 0.714 & 0 & 0 & 0 & 0 & 0 & 0 & 0 & 0 & 0 & 0 & 0 \\
0.099 & 0 & 0.156 & 0 & 0.745 & 0 & 0 & 0 & 0 & 0 & 0 & 0 & 0 & 0 & 0 \\
0 & 0.176 & 0 & 0 & 0 & 0.824 & 0 & 0 & 0 & 0 & 0 & 0 & 0 & 0 & 0 \\
0.185 & 0 & 0 & 0 & 0.074 & 0.185 & 0.556 & 0 & 0 & 0 & 0 & 0 & 0 & 0 & 0 \\
0 & 0.241 & 0 & 0.084 & 0 & 0.133 & 0 & 0.542 & 0 & 0 & 0 & 0 & 0 & 0 & 0 \\
0 & 0 & 0.286 & 0 & 0.143 & 0 & 0 & 0 & 0.571 & 0 & 0 & 0 & 0 & 0 & 0 \\
0 & 0 & 0 & 0.339 & 0 & 0 & 0 & 0.068 & 0.169 & 0.424 & 0 & 0 & 0 & 0 & 0 \\
0 & 0 & 0 & 0 & 0.419 & 0 & 0.073 & 0 & 0.115 & 0 & 0.393 & 0 & 0 & 0 & 0 \\
0 & 0 & 0 & 0 & 0 & 0.48 & 0 & 0.12 & 0 & 0 & 0 & 0.4 & 0 & 0 & 0 \\
0 & 0 & 0 & 0 & 0 & 0 & 0.37 & 0 & 0 & 0 & 0.074 & 0.185 & 0.37 & 0 & 0 \\
0 & 0 & 0 & 0 & 0 & 0 & 0 & 0.455 & 0 & 0.08 & 0 & 0.125 & 0 & 0.341 & 0 \\
0 & 0 & 0 & 0 & 0 & 0 & 0 & 0 & 0.522 & 0 & 0.13 & 0 & 0 & 0 & 0.348 \\
0.482 & 0.225 & 0.292 & 0 & 0 & 0 & 0 & 0 & 0 & 0 & 0 & 0 & 0 & 0 & 0 \\
0.173 & 0.499 & 0.328 & 0 & 0 & 0 & 0 & 0 & 0 & 0 & 0 & 0 & 0 & 0 & 0 \\
0.106 & 0.372 & 0.522 & 0 & 0 & 0 & 0 & 0 & 0 & 0 & 0 & 0 & 0 & 0 & 0
\end{pmatrix}$$

Fig. 9.11. Matrix Q.

Table 9.3. Mean sojourn times $ET(i,x)$.

	$x=0$	$x=1$	$x=2$	$x=3$	$x=4$
$i=0$	0.204	0.227	0.256	0.294	0.997
$i=1$	0.142	0.159	0..180	0.208	0.903
$i=2$	0.118	0.133	0.154	0.182	0.886

Table 9.4. The stationary probabilities of states (9.19).

	$x=0$	$x=1$	$x=2$	$x=3$	$x=4$
$i=0$	0.031	0.048	0.042	0.025	0.049
$i=1$	0.061	0.096	0.076	0.040	0.109
$i=2$	0.063	0.101	0.078	0.040	0.141

probabilities of the states are as follows:

$$q := (0.024 \quad 0.067 \quad 0.083 \quad 0.04 \quad 0.123 \quad 0.165 \quad 0.038 \quad 0.113 \quad 0.15 \quad 0.021 \quad 0.055 \quad 0.071 \quad 7.588 \times 10^{-3} \quad 0.019 \quad 0.025)$$

Table 9.3 presents the mean sojourn times $ET(i,x)$ in states (i,x) calculated using formulas (9.15) and (9.16).

The stationary probabilities of states are presented in Table 9.4.

Table 9.5. Mean incomes $E(In(s))$.

	$s=0$	$s=1$	$s=2$	$s=3$	$s=4$
R	0.516	0.620	0.701	0.734	0.734
$c=0.2$	3.613	4.138	4.506	**4.540**	4.340
$c=0.3$	3.613	4.036	**4.306**	4.240	3.940

The stationary probability R that the system is worked equals 0.701.

We finalize our example using a cost analysis. Let the income per unit time of the worked system be $r = 7$ euro, and expense for one server be $c = 0.2$ or $c = 0.3$ euro at unit time. Table 9.5 represents the mean income $E(In(s))$ as a function of the server number s.

The table shows that the optimal number of servers is three if $c = 0.2$ and two if $c = 0.3$.

9.7 Conclusion

A popular reliability model *k-out-of-n* is considered for the case in which the system operates in an external random environment. The last is a continuous-time finite Markov chain. The state of the worked system is determined by the number of failed components. Consequently, we have a two-dimensional continuous-time Markov chain. It allows for calculating the time distribution until system failure.

Capital repairs renewed the system fully. Their durations were generally distributed. As result, a semi-Markov process occurred. The jumps in the process correspond to changes in the state of the process. The stationary probabilities of the states were calculated.

This elaborate approach can be applied to the analysis of various *k-out-of-n* systems that operate in an external random environment.

References

[1] Dimitrov, B., Rykov, V., On k-out-of-n System Under Full Repair and Arbitrary Distributed Repair Time. In: Vishnevskiy, V.M., Samouylov, K.E., Kozyrev, D.V. (eds.) *DCCN 2021. LNCS*, Vol. 13144. Springer, 2021, pp. 323–335.

[2] Eryilmaz, S., Review of recent advances in reliability of consecutive k-out-of-n and related systems. *Proceedings of the Institute of Mechanical Engineering Part O J. Risk Reliability*, 2010;224:225–237.

[3] Barron, Y., Frostig E., Levikson B., Analysis of R-out-of-N systems with several repairmen, exponential life times and phase type repair times: An algorithmic approach. *European Journal of Operational Research*, 2006;169:202–225.

[4] Hellmich, M., Semi-Markov embeddable reliability structures and applications to load-sharing k-out-of-n system. *International Journal of Reliability, Quality and Safety Engineering*, 2013;20.

[5] Rykov, V., Kozyrev, D., Filimonov, A., Ivanova, N., On reliability function of a k-out-of-n system with general repair time distribution. *Probability of Engineering Informative Science*, 2020;35:1–18.

[6] Rykov, V., Ivanova, N., Kozyrev, D., Sensitivity analysis of a k-out-of-n: F system characteristics to shapes of input distribution. In: *Lecture Notes in Computer Science (LNCS)*, Vol. 12563. 2020, pp. 485–496.

[7] Gökdere, G., Ng, H.K.T., Time-dependent reliability analysis for repairable consecutive-k-out-of-n: F system. *Statistical Theory Relative Fields*, 2022;6:139–147.

[8] Khatab, A., Nahas, N., Nourelfath, M., Availability of k-out-of-n: G systems with non-identical components subject to repair priorities. *Reliability Engineering and System Safety*, 2009;94(2):142–151.

[9] Sutar, S., Naik-Nimbalkar, U.V., A load share model for non-identical components of a k-out-of-m system. *Applied Mathematics Modeling*, 2019;72:486–498.

[10] Limnios, N., Oprisan, G., *Semi-Markov Processes and Reliability.* Birkhauser, Boston, 2001.

[11] Rykov, V., Ivanova, N., Kozyrev, D., Application of decomposable semiregenerative processes to the study of k-out-of-n systems. *Mathematics*, 2021;9:1933.

[12] Wu, Y., Guan, J., Repairable consecutive-k-out-of-n: G systems with r repairmen. *IEEE Transactions of Reliability*, 2005;54:328–337.

[13] Ragi Krishnan, Somasundaram, S., The reliability and profit analysis of k-out-of-n: G systems with sensor. *European Journal of Scientific Research*, 2012;67(2):215–222.

[14] Sleeper, A.D., *Six Sigma Distribution Modelling.* McGraw-Hill Co., New York, 2006.

[15] Gertsbakh, I., Shpungin, Y., Reliability of heterogeneous ((k,r)-out-of-(n, m)) system. *RT&A*, 2016;11:8–10.

[16] Houankpo, H. G. K., Kozyrev, D., Reliability model of a homogeneous hot-standby k-Out-of-n: G system. In: Vishnevskiy, V.M., Samouylov, K.E., Kozyrev, D.V. (eds.) *DCCN 2021.* LNCS, Vol. 13144. Springer, 2021, pp. 358–368.

[17] Liang, X., Xiong, Y., Li, Z., Exact reliability formula for consecutive-k-out-of-n repairable systems. *IEEE Transaction Reliability*, 2010;59(2):313–318.

[18] Rykov, V., Kochueva, O., Rykov, Y., Preventive maintenance of k-out-of-n system with respect to cost-type criterion. *Mathematics*, 2021;9.

[19] Rykov, V.V., Decomposable semi-regenerative processes: review of theory and applications to queueing and reliability systems. *RT&A*, 2021;16(2(62)):157–190.

Chapter 10

Statistical Inferences in the Case of a Random Environment Existence

10.1 Introduction

The previous chapters raise curiosity about determining the parameters of the finite continuous-time Markov chain, which is fundamental in the exposition. In this chapter, we suggest a simple approach for the estimation of parameters and a detailed analysis of their statistical properties, which is essential because the finite continuous-time Markov chains (see Section 1.2) have the enormous applications in the theoretical and applied stochastic investigations [1–5]. Furthermore, the statistical aspects of the corresponding probabilistic models have been under-investigated [6–12].

The parameters of Markov chain are represented by the matrix $\lambda = (\lambda_{i,j}), i, j = 1, \ldots, k$, where k is number of states. We assume n observations, each corresponding to a jump in the chain. For each observation, a ternary sequence (i, j, t) is fixed, where i and j are the initial and subsequent states of the chain, respectively, and t is the time elapsed in the initial state. This case arises when all the trajectories of a considered Markov chain are fully observed.

These observations can be divided into k subsamples based on the chain's initial state. Therefore, the problem reduces to estimation parameters $\lambda_{i,j}$ for fixed i based on samples (j_r, t_r), $r = 1, 2, \ldots, n_i$. Let $N_{i,,j}$ be the number of occurrences of state j in subsamples (j_r, t_r), $r = 1, 2, \ldots, n_i$. Then, subsamples

$$n_i = N_{i,1} + N_{i,2} + \cdots + N_{i,k}. \tag{10.1}$$

Let $\Lambda_i = \lambda_{i,1} + \lambda_{i,2} + \cdots + \lambda_{i,k}$, $\mu_{i,j} = \frac{1}{\lambda_{i,j}}$, $M_i = 1/\Lambda_i$. The corresponding estimates are denoted with an asterisk: $\lambda^*_{i,j}$, Λ^*_i, $\mu^*_{i,j}$, M^*_i. We now consider several estimates of that kind, and investigate their properties for small and big samples.

10.2 Density of distribution for $\{\lambda^*_{i,j}\}$

We have,

$$M^*_i = \frac{1}{n_i}\sum_{r=1}^{n_i} t_r. \tag{10.2}$$

which provides the following estimates of interest:

$$\Lambda^*_i = \frac{1}{M*_i}, \quad \lambda^*_{i,j} = \frac{N_{i,j}}{n_i}\Lambda^*_i, \quad \mu^*_{i,j} = 1/\lambda^*_{i,j}. \tag{10.3}$$

These simple estimates are of great significance as they allow us to apply the aforementioned Markov chain models. The statistical properties of these estimates are investigated further.

It is known that the sum $\sum_{r=1}^{n_i} t_r$ has Erlang distribution with parameters Λ_i and n_i having the density

$$\tilde{f}_i(z) = \frac{1}{(n_i-1)!}\Lambda_i(z\Lambda_i)^{n_i-1}\exp(-z\Lambda_i), \quad z \geq 0. \tag{10.4}$$

Therefore, the estimate M^*_i has the following density:

$$f_i(z) = n_i\tilde{f}_i(zn_i) = \frac{n_i}{(n_i-1)!}\Lambda_i(zn_i\Lambda_i)^{n_i-1}\exp(-zn_i\Lambda_i), \quad z \geq 0. \tag{10.5}$$

Now, we can represent cumulative $P\{\Lambda^*_i \leq z\}$ and density $g_i(z)$ functions for Λ^*_i as follows:

$$\begin{aligned} P\{\Lambda^*_i \leq z\} &= P\left\{\frac{1}{M^*_i} \leq z\right\} = P\left\{M^*_i \geq \frac{1}{z}\right\} = 1 - P\left\{M^*_i \leq \frac{1}{z}\right\} \\ &= 1 - \int_0^{\frac{1}{z}} f_i(x)dx, \end{aligned}$$

$$g_i(z) = z^{-2} f_i(z^{-1}) = z^{-2} \frac{n_i}{(n_i - 1)!} \Lambda_i \left(n_i \Lambda_i \frac{1}{z} \right)^{n_i - 1}$$
$$\times \exp\left(-n_i \Lambda_i \frac{1}{z} \right), \quad z \geq 0. \tag{10.6}$$

The following inference shows that the normalization property is fulfilled:

$$\int_0^\infty g_i(z)dz = \int_0^\infty z^{-2} \frac{n_i}{(n_i - 1)!} \Lambda_i \left(n_i \Lambda_i \frac{1}{z} \right)^{n_i - 1} \exp\left(-n_i \Lambda_i \frac{1}{z} \right) dz$$
$$= -\int_0^\infty \frac{n_i}{(n_i - 1)!} \Lambda_i \left(n_i \Lambda_i \frac{1}{z} \right)^{n_i - 1} \exp\left(-n_i \Lambda_i \frac{1}{z} \right) d\left(\frac{1}{z} \right)$$
$$= \int_0^\infty \frac{n_i \Lambda_i}{(n_i - 1)!} (n_i \Lambda_i x)^{n_i - 1} \exp(-n_i \Lambda_i x) d(x) = 1.$$

We calculate the rth moment of estimate Λ_i^* for $r \leq n_i$ as follows:

$$E(\Lambda_i^{*r}) = \int_0^\infty g_i(x) x^r dx = \int_0^\infty \frac{n_i \Lambda_i}{(n_i - 1)!} (n_i \Lambda_i x)^{n_i - 1}$$
$$\times \exp(-n_i \Lambda_i x) \left(\frac{1}{x} \right)^r dx$$
$$= \prod_{j=1}^r \frac{1}{(n_i - j)!} (n_i \Lambda_i)^r \int_0^\infty \frac{n_i \Lambda_i}{(n_i - r - 1)!} (n_i \Lambda_i x)^{(n_i - r) - 1}$$
$$\times \exp(-n_i \Lambda_i x) dx$$
$$= \prod_{j=1}^r \frac{1}{(n_i - j)!} (n_i \Lambda_i)^r.$$

Therefore,

$$E(\Lambda_i^{*r}) = (n_i \Lambda_i)^r \prod_{j=1}^r \frac{1}{(n_i - j)!}. \tag{10.7}$$

Particularly,

$$E(\Lambda_i^*) = \frac{n_i}{n_i - 1} \Lambda_i, \quad E((\Lambda_i^*)^2) = \frac{n_i^2}{(n_i - 1)(n_i - 2)} \Lambda_i^2,$$

$$\mathrm{Var}(\Lambda_i^*) = \frac{n_i^2}{(n_i - 1)^2 (n_i - 2)} \Lambda_i^2. \tag{10.8}$$

Random vector $(N_{i,1}, N_{i,2}, \dots, N_{i,k})$ has polynomial distribution with parameters n_i and $\frac{\lambda_{i,1}}{\Lambda_i}, \frac{\lambda_{i,2}}{\Lambda_i}, \dots, \frac{\lambda_{i,k}}{\Lambda_i}$:

$$\begin{aligned} P\{N_{i,1} &= n_{i,1}, N_{i,2} = n_{i,2}, \dots, N_{i,k} = n_{i,k}\} \\ &= \frac{n_i!}{n_{i,1}!n_{i,2}!\cdots n_{i,k}!}\left(\frac{\lambda_{i,1}}{\Lambda_i}\right)^{n_{i,1}}\left(\frac{\lambda_{i,2}}{\Lambda_i}\right)^{n_{i,2}}\cdots\left(\frac{\lambda_{i,k}}{\Lambda_k}\right)^{n_{i,k}} \\ &= \frac{n_i!}{n_{i,1}!n_{i,2}!\cdots n_{i,k}!}\left(\frac{M_1}{\mu_{i,1}}\right)^{n_{i,1}}\left(\frac{M_2}{\mu_{i,2}}\right)^{n_{i,2}}\cdots\left(\frac{M_k}{\mu_{i,k}}\right)^{n_{i,k}}, \\ n_{i,1} &+ \cdots + n_{i,k} = n_i, \quad i = 1, \dots, k. \end{aligned} \tag{10.9}$$

Here, $n_{i,1}, n_{i,2}, \dots, n_{i,k}$ are natural numbers from the set $\{0, 1, \dots, n_i\}$, the sum of which is n_i. Let us denote $\Omega(n_i, k)$ as the set of all such vectors $\tilde{n}_i = (n_{i,1}, n_{i,2}, \dots, n_{i,k})$.

Note that partial variable $N_{i,j}$ has binomial distribution with parameters n_i and $\lambda_{i,j}/\Lambda_i$. Therefore, the expectation and variance are calculated as follows:

$$E(N_{i,j}) = n_i\frac{\lambda_{i,j}}{\Lambda_i}, \quad \text{Var}(N_{i,j}) = n_i\frac{\lambda_{i,j}}{\Lambda_i}\left(1 - \frac{\lambda_{i,j}}{\Lambda_i}\right).$$

Therefore, the estimate $\lambda^*_{i,j}$ (see formula (10.3)) has the following density:

$$h_{i,j}(z) = \sum_{\eta=1}^{n_i}\binom{n_i}{\eta}\left(\frac{\lambda_{i,j}}{\Lambda_i}\right)^{\eta}\left(1 - \frac{\lambda_{i,j}}{\Lambda_i}\right)^{n_i-\eta}\frac{n_i}{\eta}g\left(z\frac{n_i}{\eta}\right), \quad z \geq 0. \tag{10.10}$$

This distribution has the following singular component:

$$P\{\lambda^*_{i,j} = 0\} = \left(1 - \frac{\lambda_{i,j}}{\Lambda_i}\right)^{n_i}. \tag{10.11}$$

It is evident, that $\mu^*_{i,j} = \infty$ when $\lambda^*_{i,j} = 0$. Therefore, the distribution of $\mu^*_{i,j}$ is singular.

Formulas (10.6) and (10.10) perform poorly for a large sample size. In this case, the normal approximation of Erlang (10.4) and binomial distributions can be used.

In the case of an approximation of density function (10.6), an observation for the initial state i gives a result with expectation

$1/\Lambda_i$ and variance $1/\Lambda_i^2$. The empirical mean (10.2) has the same expectation and variance $1/n_i\Lambda_i^2$. Therefore, density (10.6) can be approximated using the normal density $fn_i(\ldots)$:

$$\tilde{g}_i(z) = z^{-2} fn_i(z^{-1}) = z^{-2}\Lambda_i\sqrt{\frac{n_i}{2\pi}}\exp\left(-\frac{1}{2}\left(\Lambda_i\sqrt{n_i}\left(\frac{1}{z} - 1/\Lambda_i\right)\right)^2\right)$$

$$= z^{-2}\Lambda_i\sqrt{\frac{n_i}{2\pi}}\exp\left(-z^{-2}\frac{1}{2}(\Lambda_i\sqrt{n_i}(1 - z/\Lambda_i))^2\right), \quad z \in (-\infty, \infty). \tag{10.12}$$

The density $\tilde{h}_{i,j}(z)$ of estimation $\lambda_{i,j}^*$ can be approximated by (10.10), where $\tilde{g}_i(z)$ must be used instead of $g_i(z)$.

The binomial distribution used in formula (10.10) is approximated using normal distribution with mean $n_i\frac{\lambda_{i,j}}{\Lambda_i}$ and variance $n_i\frac{\lambda_{i,j}}{\Lambda_i}\left(1 - \frac{\lambda_{i,j}}{\Lambda_i}\right)$. Let $dnorm\left(u, n_i\frac{\lambda_{i,j}}{\Lambda_i}, \sqrt{n_i\frac{\lambda_{i,j}}{\Lambda_i}\left(1 - \frac{\lambda_{i,j}}{\Lambda_i}\right)}\right)$ be the corresponding density function. Now, instead of densities $h_{i,j}(z)$ and $\tilde{h}_{i,j}(z)$, we have the following density for $-\infty \leq z \leq \infty$:

$$\tilde{\tilde{h}}_{i,j}(z) = \int_0^n \frac{n_i}{u}\tilde{g}\left(z\frac{n_i}{u}\right)\text{dnorm}\left(u, n_i\frac{\lambda_{i,j}}{\Lambda_i}, \sqrt{n_i\frac{\lambda_{i,j}}{\Lambda_i}\left(1 - \frac{\lambda_{i,j}}{\Lambda_i}\right)}\right)du$$

$$= \int_0^n \frac{n_i}{u}\left(z\frac{n_i}{u}\right)^{-2}\Lambda_i\sqrt{\frac{n_i}{2\pi}}$$

$$\times \exp\left(-\left(z\frac{n_i}{u}\right)^{-2}\frac{1}{2}\left(\Lambda_i\sqrt{n_i}\left(1 - \frac{zn_i}{u\Lambda_i}\right)\right)^2\right)$$

$$\times \text{dnorm}\left(u, n_i\frac{\lambda_{i,j}}{\Lambda_i}, \sqrt{n_i\frac{\lambda_{i,j}}{\Lambda_i}\left(1 - \frac{\lambda_{i,j}}{\Lambda_i}\right)}\right)du. \tag{10.13}$$

10.3 Expectation, variance and covariance for $\{\lambda_{i,j}^*\}$

Densities $h_{i,j}(z), \tilde{h}_{i,j}(z)$, and $\tilde{\tilde{h}}_{i,j}$ facilitate the calculation of expectation and variance of the estimate $\lambda_{i,j}^*$. Moments of the estimate $\lambda_{i,j}^*$

can be calculated as follows:

$$E({\lambda^*_{i,j}}^r) = E\left(\left(\frac{N_{i,j}}{n_i}\right)^r\right) E(\Lambda_i^{*r}), \quad r = 1, 2, \ldots \tag{10.14}$$

Particularly,

$$E(\lambda^*_{i,j}) = \frac{\lambda_{i,j}}{\Lambda_i} E(\Lambda_i^*) = \frac{\lambda_{i,j}}{\Lambda_i} \frac{n_i}{n_i - 1}\Lambda_i = \frac{n_i}{n_i - 1}\lambda_{i,j}, \tag{10.15}$$

$$\begin{aligned}
E({\lambda^*_{i,j}}^2) &= E\left(\left(\frac{N_{i,j}}{n_i}\right)^2\right) E(\Lambda_i^{*2}) \\
&= \left(\left(\frac{\lambda_{i,j}}{\Lambda_i}\right)^2 + \frac{1}{n_i}\frac{\lambda_{i,j}}{\Lambda_i}\left(1 - \frac{\lambda_{i,j}}{\Lambda_i}\right)\right) \frac{n_i^2}{(n_i-1)(n_i-2)}\Lambda_i^2 \\
&= \left(\lambda_{i,j}^2 + \frac{1}{n_i}\lambda_{i,j}(\Lambda_i - \lambda_{i,j})\right) \frac{n_i^2}{(n_i-1)(n_i-2)}, \\
\mathrm{Var}(\lambda^*_{i,j}) &= E({\lambda^*_{i,j}}^2) - (E(\lambda^*_{i,j}))^2 \\
&= \left(\lambda_{i,j}^2 + \frac{1}{n_i}\lambda_{i,j}(\Lambda_i - \lambda_{i,j})\right) \frac{n_i^2}{(n_i-1)(n_i-2)} - \left(\frac{n_i}{n_i-1}\lambda_{i,j}\right)^2 \\
&= \lambda_{i,j}^2 \frac{n_i^2}{(n_i-1)^2(n_i-2)} + \lambda_{i,j}(\Lambda_i - \lambda_{i,j}) \frac{n_i}{(n_i-1)(n_i-2)} \\
&= \lambda_{i,j}^2 \frac{n_i}{(n_i-1)^2(n_i-2)} + \lambda_{i,j}\Lambda_i \frac{n_i}{(n_i-1)(n_i-2)}.
\end{aligned} \tag{10.16}$$

Now, we calculate a covariance between parameters $\lambda^*_{i,j}$ and $\lambda^*_{i,j*}$ for differents j and j^*. The second mixed moment for $N_{i,j}$ and $N_{i,j*}$ is given as follows:

$$\begin{aligned}
E(N_{i,j}N_{i,j*}) &= \sum_{n_{i,j}=0}^{n_i} P\{N_{i,j} = n_{i,j}\} n_{i,j} E(N_{i,j*} | N_{i,j} = n_{i,j}) \\
&= \sum_{n_{i,j}=0}^{n_i} \frac{n_i!}{n_{i,j}!(n_i - n_{i,j})!} \left(\frac{\lambda_{i,j}}{\Lambda_i}\right)^{n_{i,j}} \left(1 - \frac{\lambda_{i,j}}{\Lambda_i}\right)^{n_i - n_{i,j}} \\
&\quad \times n_{i,j}(n_i - n_{i,j}) \frac{\lambda_{i,j*}}{\Lambda_i - \lambda_{i,j}}
\end{aligned}$$

$$= \frac{\lambda_{i,j*}}{\Lambda_i - \lambda_{i,j}} \left(n_i^2 \frac{\lambda_{i,j}}{\Lambda_i} - \left(\left(n_i \frac{\lambda_{i,j}}{\Lambda_i} \right)^2 + n_i \frac{\lambda_{i,j}}{\Lambda_i} \left(1 - \frac{\lambda_{i,j}}{\Lambda_i} \right) \right) \right)$$

$$= n_i(n_i - 1) \frac{\lambda_{i,j}\lambda_{i,j*}}{\Lambda_i^2} \tag{10.17}$$

This result allows to produce the second mixed moment and the covariance for $\lambda_{i,j}^*$ and $\lambda_{i,j*}^*$ as follows:

$$E(\lambda_{i,j}^* \, \lambda_{i,j*}^*) = E\left(\frac{N_{i,j}}{n_i} \Lambda_i^* \frac{N_{i,j*}}{n_i} \Lambda_i^* \right) = n_i^{-2} E(N_{i,j} N_{i,j*}) E(\Lambda_i^{*2})$$

$$= \frac{(n_i - 1)}{n_i} \frac{\lambda_{i,j}\lambda_{i,j*}}{\Lambda_i^2} \frac{n_i^2}{(n_i - 1)(n_i - 2)} \Lambda_i^2$$

$$= \frac{n_i}{n_i - 2} \lambda_{i,j}\lambda_{i,j*}, \tag{10.18}$$

$$Cov(\lambda_{i,j}^*, \lambda_{i,j*}^*) = E(\lambda_{i,j}^* \, \lambda_{i,j*}^*) - E(\lambda_{i,j}^*) E(\lambda_{i,j*}^*)$$

$$= \frac{n_i}{n_i - 2} \lambda_{i,j}\lambda_{i,j*} - \frac{n_i}{n_i - 1} \lambda_{i,j} \frac{n_i}{n_i - 1} \lambda_{i,j*}$$

$$= \lambda_{i,j}\lambda_{i,j*} \frac{n_i}{(n_i - 1)^2 (n_i - 2)}. \tag{10.19}$$

10.4 Joint distribution of estimates $\{\lambda_{i,j}^*\}$

Now, we consider the distribution function of the vector $\lambda_i^* = (\lambda_{i,1}^*, \ldots, \lambda_{i,k}^*)$:

$$F_i(x) = P(\lambda_{i,1}^* \le x_1, \lambda_{i,2}^* \le x_2, \ldots, \lambda_{i,k}^* \le x_k), x = (x_1, x_2, \ldots, x_k) \ge 0.$$

Let $N_i = (N_{i,1}, N_{i,2}, \ldots, N_{i,k})$ be a vector having the distribution (10.9), and $\tilde{n}_i = (n_{i,1}, n_{i,2}, \ldots, n_{i,k}) \in \Omega(n_i, k)$ be a possible value of N_i. Then,

$$F_i(x) = \sum_{\tilde{n}_i \in \Omega(n_i,k)} P\{N_i = \tilde{n}_i\} P \left\{ \prod_{j \in \{1,2,\ldots,k\}} \{\lambda_{i,j}^* \le x_j | N_{i,j} = n_{i,j}\} \right\}$$

$$= \sum_{\tilde{n}_i \in \Omega(n_i,k)} P\{N_i = \tilde{n}_i\} P\left\{ \prod_{j \in \{1,2,\ldots,k\}} \left\{ \frac{n_{i,j}}{n_i} \Lambda_i^* \le x_j \right\} \right\}$$

$$= \sum_{\tilde{n}_i \in \Omega(n_i,k)} P\{N_i = \tilde{n}_i\} P\left\{ \prod_{j \in \{1,2,\ldots,k\}} \left\{ \Lambda_i^* \le x_j \frac{n_i}{n_{i,j}} \right\} \right\}.$$

If $\varphi(x, \tilde{n}_i) = \min_{j \in \{1,2,\ldots,k\}} x_j \frac{n_i}{n_{i,j}}$, then the following is the necessary formula:

$$F_i(x) = \sum_{\tilde{n}_i \in \Omega(n_i,k)} P\{N_i = \tilde{n}_i\} G_i(\varphi(x, \tilde{n}_i)), x = (x_1, x_2, \ldots, x_k) \ge 0, \tag{10.20}$$

where G_i is a cumulative distribution function for the density (10.6).

10.5 Asymptotic distributions

The well-known results of *large sample theory* can be used for large sample sizes [11,13–16]. These results state that the vector $\frac{1}{n_i}(N_{i,1}, N_{i,2}, \ldots, N_{i,k})$ and scalar $\Lambda*_i$, are consistently and asymptotically the unbiased estimates of $\frac{1}{\Lambda_i}(\lambda_{i,1}\lambda_{i,2}, \ldots, \lambda_{i,k})$ and $\Lambda_{i,}$.

The same result is derived using (10.15) and (10.16). Now, we prove (following [13–15]) that $\lambda_i^* = (\lambda_{i,1}^*, \ldots, \lambda_{i,k}^*,)$ has the asymptotically multivariate normal distribution.

Let $p_j = \frac{\lambda_{i,j}}{\Lambda_i}$, $j = 1, \ldots, k$, and

$$V_i^T = \begin{pmatrix} \frac{N_{i,1} - n_i p_1}{\sqrt{n_i p_1}} & \cdots & \frac{N_{i,k} - n_i p_k}{\sqrt{n_i p_k}} \end{pmatrix} = \begin{pmatrix} \frac{\frac{1}{n_i} N_{i,1} - p_1}{\sqrt{\frac{1}{n_i} p_1}} & \cdots & \frac{\frac{1}{n_i} N_{i,k} - p_k}{\sqrt{\frac{1}{n_i} p_k}} \end{pmatrix},$$

$$\varphi_{\mathrm{i}} = \begin{pmatrix} \sqrt{p_1} & \cdots & \sqrt{p_k} \end{pmatrix}^{\mathrm{T}}. \tag{10.21}$$

In accordance with (formula 6a.1.5 in [15]), the asymptotic distribution of vector V is a multivariate normal distribution with zero mean and covariance matrix

$$\mathrm{Cov}(V_i) = I - \varphi_i {\varphi_i}^T, \tag{10.22}$$

where I is a unit k-dimension matrix.

Therefore, for $j \neq j*$,

$$\operatorname{Cov}\left(\frac{1}{n_i}N_{i,j}, \frac{1}{n_i}N_{i,j*}\right) = \sqrt{\frac{1}{n_i}p_j}\sqrt{\frac{1}{n_i}p_{j*}}$$

$$\times E\left(\frac{\frac{1}{n_i}N_{i,j} - p_j}{\sqrt{\frac{1}{n_i}p_j}} \frac{\frac{1}{n_i}N_{i,j*} - p_{j*}}{\sqrt{\frac{1}{n_i}p_{j*}}}\right) + o\left(\frac{1}{n_i}\right)$$

$$= \sqrt{\frac{1}{n_i}p_j}\sqrt{\frac{1}{n_i}p_{j*}}(-1)\sqrt{p_j}\sqrt{p_{j*}} + o\left(\frac{1}{n_i}\right)$$

$$= -\frac{1}{n_i}p_j p_{j*} + o\left(\frac{1}{n_i}\right), \tag{10.23}$$

$$\operatorname{Var}\left(\frac{1}{n_i}N_{i,j}\right) = \frac{1}{n_i}p_j E\left(\frac{\frac{1}{n_i}N_{i,j} - p_j}{\sqrt{\frac{1}{n_i}p_j}}\right)^2$$

$$+o\left(\frac{1}{n_i}\right) = \frac{1}{n_i}p_j(1-p_j) + o\left(\frac{1}{n_i}\right). \tag{10.24}$$

that corresponds to (10.17).

Considering the asymptotical distribution of estimate $\Lambda_i^* = \frac{1}{M_i^*}$, we know that the random variable $\sqrt{n_i}\left(M_i^* - \frac{1}{\Lambda_i}\right)$ has asymptotically normal distribution with zero mean and variance $\sigma^2(\Lambda_i) = 1/\Lambda_i^2$. Based on the formula (6a.2.1 in [15]), $\tilde{\Lambda} = \sqrt{n_i}(\Lambda_i^* - \Lambda_i)$ has asymptotically normal distribution with zero mean and variance

$$\operatorname{Var}(\tilde{\Lambda}) = \sigma^2(\Lambda_i)\left(\frac{\partial}{\partial \propto}\frac{1}{\propto}\right)^2 |_{\propto=\frac{1}{\Lambda_i}} = (1/\Lambda_i^2)\Lambda_i^4 = \Lambda_i^2. \tag{10.25}$$

Consequently, asymptotical variance of $\Lambda*_i$ equals $\frac{1}{n_i}\Lambda_i^2 + o\left(\frac{1}{n_i}\right)$. From (10.15) and (10.16) we obtain the following:

$$E(\lambda_{i,j}^*) = \lambda_{i,j} + o\left(\frac{1}{n_i}\right),$$

$$\operatorname{Var}(\lambda_{i,j}^*) = \frac{1}{n_i}\lambda_{i,j}\Lambda_i + o\left(\frac{1}{n_i}\right).$$

Finally, the asymptotical distribution of estimates $\lambda_i^* = (\lambda_{i,1}^*, \ldots, \lambda_{i,k}^*,)$ is considered.

The $(k+1)$–vector $U_i = \left(\Lambda_i^* \quad \frac{1}{n_i}N_{i,1}\cdots\frac{1}{n_i}N_{i,k}\right)^T$ has independent components Λ_i^* and $\left(\frac{1}{n_i}N_{i,1}\cdots\frac{1}{n_i}N_{i,k}\right)$, and the scalar $(\Lambda_i^* - \Lambda_i)\sqrt{n_i}$ has asymptotically normal distribution with zero mean and variance Λ_i^2. The k-vector $\left(\frac{1}{n_i}N_{i,1} - p_1 \quad \cdots \quad \frac{1}{n_i}N_{i,k} - p_k\right)\sqrt{n_i}$ has asymptotically multivariate normal distribution with zero mean and covariance matrix $\mathrm{n}_i \times \mathrm{Cov}$, where the matrix Cov is defined by (10.23) and (10.24).

Therefore, the vector

$$\sqrt{n_i}(U_i - E(U_i)) = \sqrt{n_i}\left(\left(\Lambda_i^* \quad \frac{1}{n_i}N_{i,1}\cdots\frac{1}{n_i}N_{i,k}\right)^T - \left(\Lambda_i \quad p_{i,1}\cdots p_{i,k}\right)^T\right)$$

has asymptotically multivariate normal distribution with zero mean and covariance matrix

$$\Omega = \begin{pmatrix} \Lambda_i^2 & 0 \\ 0 & n_i\mathrm{Cov} \end{pmatrix}. \tag{10.26}$$

Let $h_1(u), \ldots, h_k(u)$ be the functions of the real $(k+1)$-vector u with a continuous partial derivative in the neighborhood of point u_0. Furthermore, let $\nabla h_j(u_0)$ be a gradient–column of $h_j(u)$ in the point u_0, and $\nabla h(u_0)$ be $k \times (k+1)$-matrix of the gradients:

$$\nabla h(u_0)^{\boldsymbol{T}} = \left(\nabla h_1(u_0) \cdots \nabla h_{\boldsymbol{k}}(u_0)\right).$$

With respect to the *theorem on differentiation transformation* (Theorem 2.6 in [13] and Theorem 3.1.3 in [14]), the vector

$$\sqrt{n_i}(h_1(U_1) - h_1(\Lambda_i \quad p_{i,1}\cdots p_{i,k})^T, \ldots, h_k(U_k) - h_k(\Lambda_i \quad p_{i,1}\cdots p_{i,k})^T)$$

has asymptotically multivariate normal distribution with zero mean and covariance matrix

$$\Sigma = \nabla h(u_0)\boldsymbol{\Omega}\nabla h(u_0)^{\boldsymbol{T}}.$$

In our case, $u_0 = (\Lambda_i \ p_1 \cdots p_k)^T$; $h_j(u_0) = \Lambda_i p_j, j = 1, \ldots, k$; $\nabla h_j(u_0) = (p_j\ 0\cdots 0\ \Lambda_i\ 0\cdots 0)^{\boldsymbol{T}}$. Therefore,

$$\Sigma = \left(\nabla h_1(u_0) \quad \cdots \quad \nabla h_{\boldsymbol{k}}(u_0)\right)^{\boldsymbol{T}} \begin{pmatrix} \Lambda_i^2 & 0 \\ 0 & n_i\mathrm{Cov} \end{pmatrix}$$

$$\cdot \left(\nabla h_1(u_0) \quad \cdots \quad \nabla h_{\boldsymbol{k}}(u_0)\right).$$

According to the formulas (10.23) and (10.24)

$$n_i \mathrm{Cov}\left(\frac{1}{n_i}N_{i,j}, \frac{1}{n_i}N_{i,j*}\right) = -p_j p_{j*} + o(1) = -\frac{\lambda_{i,j}}{\Lambda_i}\frac{\lambda_{i,j*}}{\Lambda_i} + o(1),$$

$$n_i \mathrm{Var}\left(\frac{1}{n_i}N_{i,j}\right) = p_j(1-p_j) + o(1) = \frac{\lambda_{i,j}}{\Lambda_i}\left(1 - \frac{\lambda_{i,j}}{\Lambda_i}\right) + o(1).$$

Therefore,

$$\Sigma_{j,j*} = p_j p_{j*}\Lambda_i^2 - \Lambda_i^2 \frac{\lambda_{i,j}}{\Lambda_i}\frac{\lambda_{i,j*}}{\Lambda_i} = \frac{\lambda_{i,j}}{\Lambda_i}\frac{\lambda_{i,j*}}{\Lambda_i}\Lambda_i^2 - \Lambda_i^2 \frac{\lambda_{i,j}}{\Lambda_i}\frac{\lambda_{i,j*}}{\Lambda_i} = 0,$$

$$\Sigma_{j,j} = p_j^2\Lambda_i^2 + \Lambda_i^2\frac{\lambda_{i,j}}{\Lambda_i}\left(1 - \frac{\lambda_{i,j}}{\Lambda_i}\right) = \Lambda_i\lambda_{i,j}. \tag{10.27}$$

The following theorem presents our conclusion:

Theorem. *The vector* $\sqrt{n_i}(\lambda_i^* - \lambda_i)$ *has multivariate asymptotically normal distribution with zero mean and covariance matrix* Σ*, defined by* (10.27).

10.6 Example

Let us illustrate the previous results by considering a chain with three states ($k = 3$), such that the initial state is $i = 1$ and the vector of transition intensities is $\lambda_i = (\lambda_{i,1} \;\; \lambda_{i,2}, \;\; \lambda_{i,3}) = (0.5 \;\; 1 \;\; 1.5)$. The number of observations n varies.

The intensity of a jump from the considered initial state is $\Lambda_1 = 3$. Figure 10.1 contains a probability density plot (10.6) for the empirical value Λ_1^* calculated for $n = 6$ and $n = 20$ observations, where $g\Lambda(z, 3, n) = g_1(z)$. The graphics of distribution densities (10.10) for parameter's estimates $\lambda_{1,1}^*\lambda_{1,2}^*\lambda_{1,3}^*$ are presented in Figs. 10.2–10.4, where $h\lambda(z, \lambda\mathrm{Our}, j-1, n) = h_{1,j}(z)$.

Table 10.1 contains expectations $E(\lambda_{1,1}^*)$, $E(\lambda_{1,2}^*)$, $E(\lambda_{1,3}^*)$ for various values of observation number n calculated using (10.15).

The results presented in Figs. 10.1–10.4 were calculated using formulas (10.6) and (10.10). However, these equations cannot be applied when the number of observations n exceeds the interval (65–80). In this case, we used the normal approximations (10.12) and (10.13). Figure 10.5 shows the graphics of distribution densities (10.10) and

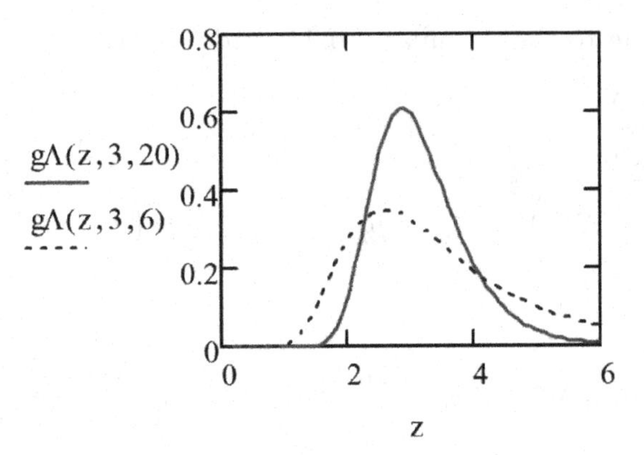

Fig. 10.1. Graphics of distribution densities (10.6) for $n = 6$ and $n = 20$.

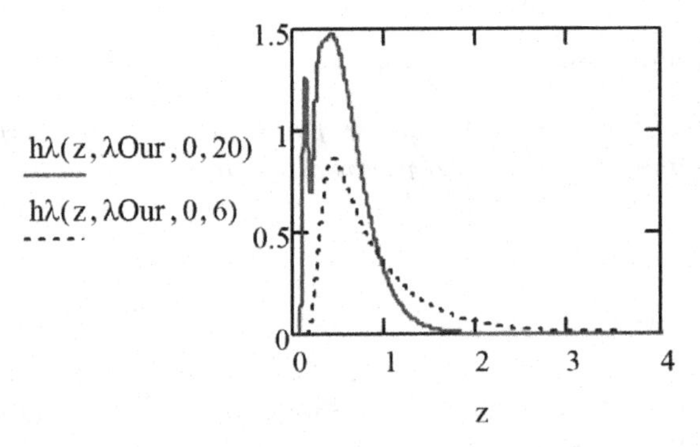

Fig. 10.2. Graphics of distribution density (10.10) for $\lambda^*_{1,1}$ and $n = 6$ and $n = 20$.

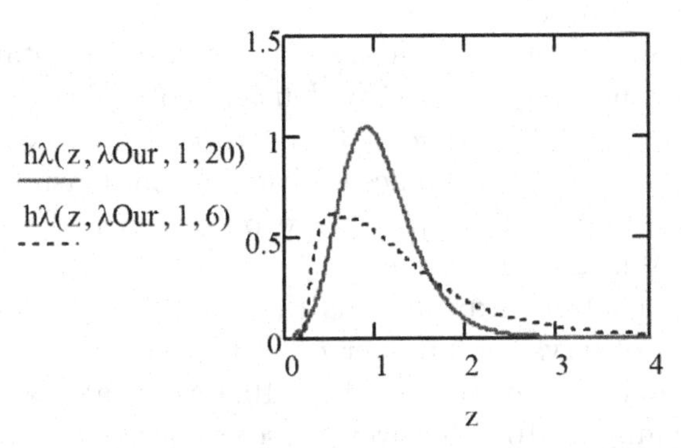

Fig. 10.3. Graphics of distribution density (10.10) for $\lambda^*_{1,2}$ and $n = 6$ and $n = 20$.

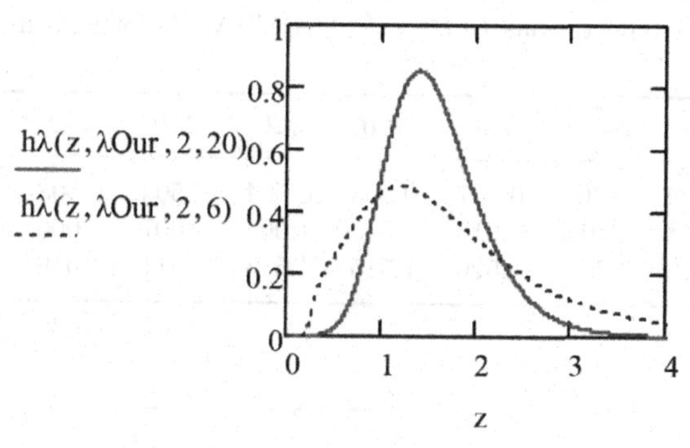

Fig. 10.4. Graphics of distribution density (10.10) for $\lambda^*_{1,3}$ and $n = 6$ and $n = 20$.

Table 10.1. Expectations $E(\lambda^*_{1,1}), E(\lambda^*_{1,2}), E(\lambda^*_{1,3})$ for various observations' number n.

n	5	10	15	20	30	40	50	60	65
$E(\lambda^*_{1,1})$	0.606	0.555	0.536	0.526	0.517	0.513	0.510	0.508	0.508
$E(\lambda^*_{1,2})$	1.159	1.107	1.071	1.053	1.034	1.026	1.020	1.017	1.016
$E(\lambda^*_{1,3})$	1.637	1.643	1.604	1.579	1.552	1.538	1.531	1.525	1.523

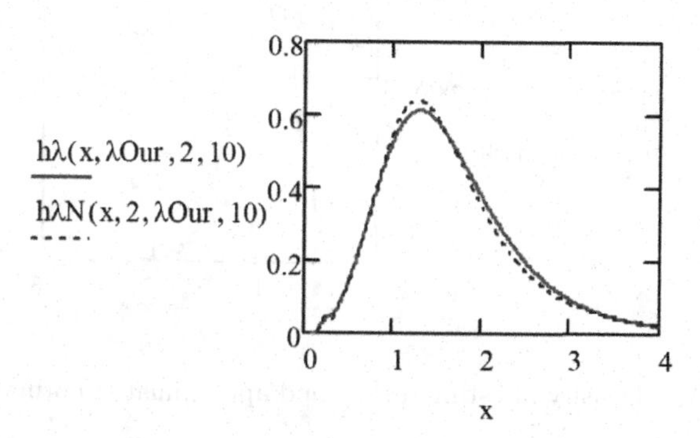

Fig. 10.5. Graphics of distribution densities (10.10) and (10.13) for $\lambda^*_{1,3}$.

(10.13) for parameter's estimate $\lambda^*_{1,3}$ and $n = 10$. It can be observed that they coincide with each other.

Table 10.2 is analogous to Table 10.1; however, the expectations are calculated using the normal approximation (10.13).

Table 10.2. Expectations $E(\lambda^*_{1,1}), E(\lambda^*_{1,2}), E(\lambda^*_{1,3}{}^*)$ for a normal approximation.

n	60	70	80	100	120	140	160	300	400
$E(\lambda^*_{1,1})$	0.509	0.507	0.507	0.505	0.504	0.504	0.503	0.502	0.501
$E(\lambda^*_{1,2})$	1.018	1.015	1.013	1.010	1.009	1.007	1.006	1.003	1.003
$E(\lambda^*_{1,3})$	1.526	1.522	1.520	1.515	1.513	1.511	1.510	1.505	1.504

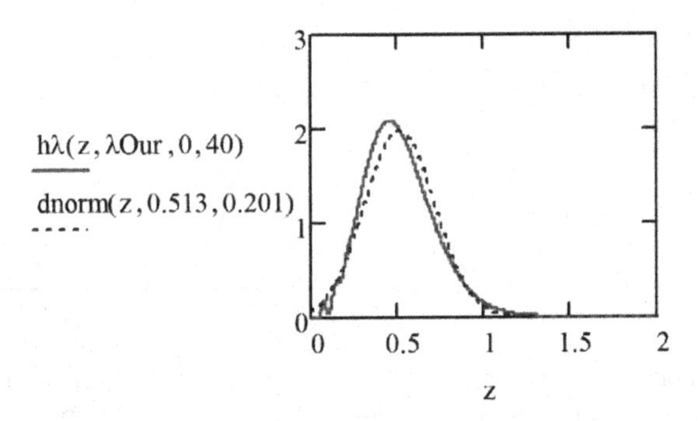

Fig. 10.6. Density of estimate λ^*_{11} and approximated normal density.

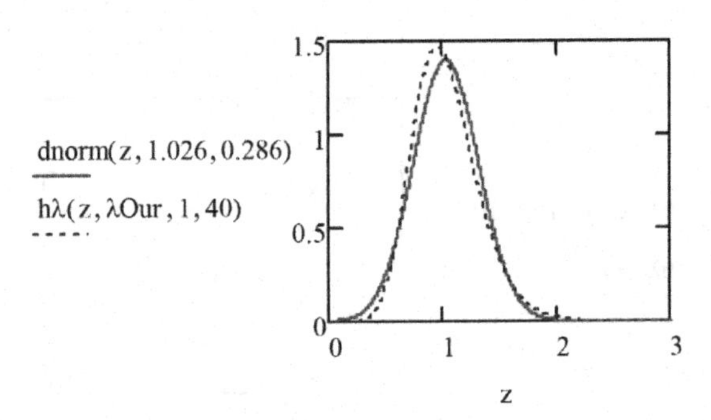

Fig. 10.7. Density of estimate $\lambda^*_{1,2}$ and approximated normal density.

For a large number of observations n, the considered densities are better approximated using the normal density. For instance, let us assume $n = 40$. Using formulas (10.15) and (10.16), the expectation, variance, and standard deviation of each estimate can be calculated as follows: $E(\lambda^*_{1,1}) = 0.513, \sigma(\lambda^*_{1,1})) = 0.201, E(\lambda^*_{1,2}) = 1.026, \sigma(\lambda^*_{1,2}) = 0.286, E(\lambda^*_{1,3}) = 1.538, \sigma(\lambda^*_{1,3}) = 0.351$.

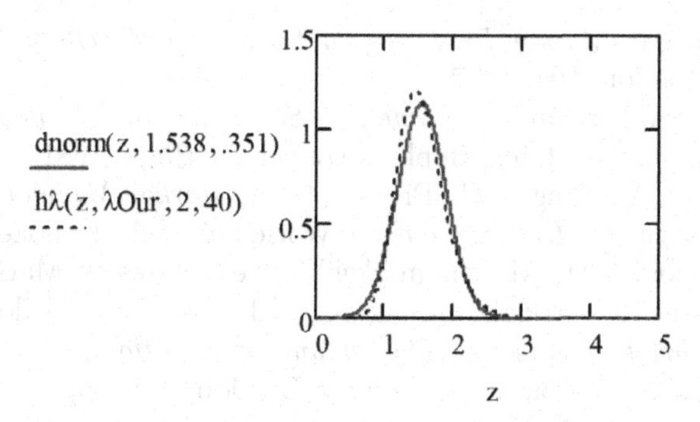

Fig. 10.8. Density of estimate $\lambda^*_{1,3}$ and approximated normal density.

Figures 10.6–10.8 depict two graphics: the densities (10.10) for considered estimates $\lambda^*_{1,1}, \lambda^*_{1,2}, \lambda^*_{1,3}$ and the density of the normal distribution *dnorm* with the calculated mean and standard deviation. Evidently, all the compared graphics coincide. If the sample size n is larger than 60, the graphics coincide completely.

10.7 Conclusion

We have considered the problem of parameter estimation for the basic probabilistic model in our monograph, and the finite continuous-time Markov chain. Furthermore, a simple estimation method was suggested. The distributions of the estimates were obtained for small and large samples, which facilitated various statistical inferences regarding the hypotheses and confidence interval of the parameters. Furthermore, it is possible to analyze the influence of sample size on the reliability of inferences about the indices that are defined for the considered Markov chain. Numerical examples were used to illustrate the properties of the proposed estimates.

References

[1] Feller, W., *An Introduction to Probability Theory and its Applications.* Vol. 2. Wiley, New York, 1971.

[2] Gnedenko, B.W., Kovalenko, I.N., *Introduction to Queueing Theory.* Nauka, Moscow, Russia, 1987 (in Russian).

[3] Kijima, M., *Markov Processes for Stochastic Modelling.* Chapman & Hall, London, UK, 1997.

[4] Neuts, M.F., *Matrix-geometric Solutions in Stochastic Models. Baltimore.* The Johns Hopkins University Press, 1981.

[5] Pacheco, A., Tang, L.C., Prabhu, N.U., *Markov-Modulated Processes & Semiregenerative Phenomena.* World Scientific, Hoboken, NJ, 2009.

[6] Andronov, A.M., Maximum likelihood estimates for Markov-additive processes of arrivals by aggregated data. In: Kollo, T. (ed.), *Multivariate Statistics: Theory and Applications,* World Scientific Publishing Company, New Jersey, London, 2013, pp. 17–33.

[7] Andronov, A.M., Markov-modulated samples and their applications. In: Melas, V.B., Mignani, S., Monari, P., Salmoso, L.S. (eds.), *Topics in Statistical Simulation.* Springer, New York, 2014, pp. 29–35.

[8] Andronov, A., On reward rate estimation for finite irreducible continue-time Markov chain. *Journal of Statistical Theory and Practice, Advance in Statistics and Simulation,* 2017;11(3), 407–417.

[9] Kleinhof, M., Paramonov, Ju., Paramonova, A., Regression model based on Markov chain theory for composite curve approximation, In: *ACTA ET Commendations Universitatis Tartuensis de Mathematuca,* Vol. 8. Tartu, 2008, pp. 143–153.

[10] Scott, S.L., Smyth, P., *The Markov Modulated Poisson Process and Markov Modulated Cascade with Application to Web Traffic Modelling,* Vol. 7. Oxford University Press, 2003, pp. 1–10.

[11] Shorack, G.R., Welner, J.A., *Empirical Processes with Applications in Statistics.* Wiley, New York, 1986.

[12] Turin, Yu., Fitting probabilistic automata via the EM algorithm. *Communications in Statistics/Stochastic Models,* 1996;12:405–424.

[13] Bandorf-Nielsen, O.E., Cox, D.R., *Asymptotic Techniques for Use in Statistics.* Chapman and Hall, London, New York, Tokyo, Melbourne, Madras, 1989.

[14] Kollo, T., von Rosen, D., *Advanced Multivariate Statistics with Matrices.* Springer, Dordrecht, 2005.

[15] Rao, C.R., *Linear Statistical Inference and Its Application.* John Wiley & Sons, Inc., New York, London, Sydney, 1966.

[16] Srivastava, M.S., Methods of Multivariate Statistics. Wiley-Interscience, New York, 2002.

Chapter 11

Review of Other Applications

In this chapter, we present a cursory examination of another application of continuous-time Markov chains in operational research. We began with a flow of homogeneous random events, such as Poisson, recurrent, and alternating flows. Currently, a popular method is *Markovian Arrival Process* (MAP), which is described as follows: an external random environment $J(t)$ is considered with the final number of states (see Chapter 1). When the process $J(t)$ transfers from state i to state j, an event is generated with probability $p_1(i, j)$. A transfer without generation occurs with the opposite probability: $p_0(i, j) = 1 - p_1(i, j)$. The *Batch Markovian Arrival Process* (BMAP) is used when the number of generated events is greater than one.

The considered flows are widely used in queueing theory, and these events are called *customers*, which are served during a random time. It is often assumed that service time has a probability *distribution of a phase type* (PH) developed by Neuts [1]. This distribution is analogous to the previous one, wherein one state is an absorbing state. The initial distribution of the states is fixed, and the service time is the time required to enter an absorbing state.

A description of *the queueing system* also includes the number of servers and discipline of the queue. A distinction must be made between queuing systems with rejections and those with waiting. The *retrial* queuing system or a system with *an orbit* is a new class of widely popular systems with waiting. Therefore, this case implies that a customer facing the busy state of all servers on arrival switches to *an orbit* and subsequently attempts again on the expiry of a random exponentially distributed time.

Various studies have been devoted to the queuing systems of $BMAP/M/s$, $BMAP$/PH/s types, and retrials [2–4].

All the enumerated systems were Markovian, and non-Markovian systems were typically investigated using the embedded Markov chains or semi-regenerative approaches. Initially, the classical single-server queuing system $M/G/1$ was considered for random environments [5–7], which is described as a continuous-time finite irreducible Markov chain $J(t)$. Arrived claims belong to different types determined by the state of chain $J(t)$ at the arrival epoch. Each customer type has its own inter-arrival time and service time distributions. Queue discipline is based on the first-come, first-served principle. The distributions of the waiting times for the transient and steady-state cases were obtained.

Studies have also been conducted on Markov-modulated queuing systems [8–10].

Probabilistic models of *the reliability* are another important domain of research [11–13]. The following problems are often considered: the system consisted of n channels connected in parallel, each with its own repair facility. The system operates in random environment $J(t)$ with k states. Herein, the functioning of all channels can be described using continuous time-varying processes $X_1(t), \ldots, X_n(t)$; $X_j(t) \in \{0, 1\}$, $j = 1, \ldots, n$. Supposedly, if the random environment has state i, then the sojourn time in the state 0 (working state) of all channels has exponential distribution with parameters $\mu_{1,i}, \ldots, \mu_{n,i}$. The sojourn times in state 1 (failure state) have general absolute continuous distributions with probability density functions (p.d.f.) $\alpha_1(t), \ldots, \alpha_n(t)$. These sojourn times are independent when the state of the random environment is fixed. The system operates at time t when at least one channel operates. The reliability of systems consisting of two or three elements has been previously investigated [14,15].

A generalization consists of cases in which a system fails, and at least k of its elements have failed (see [16] and the bibliography therein). Studies [17,18] analyzed the reliability when the failed elements and the entire system were repaired using a single server. However, a random environment was not considered in this study.

An important and interesting logistics problem is the reliability of *supply chains* [19–21] comprising several sequential operations. Each operation can either be fulfilled successfully or fail. It is necesary to determinate a probability of all operations being performed successfully within a fixed time. Reference [19] considers a case in which the probabilities of failure and operation duration are dependent on the state of the external random environment. The latter is described using a continuous-time finite Markov chain with known parameters. The peculiarity of the problem consists of the following: a fatal failure occurs with intensity depending on time, thereby yielding a nonhomogeneous system of differential equations that are numerically solved.

An *overbooking problem* assumes that the sale and booking of some products or services exceed the given possibilities, and it considers that a part of the sale or booking will be canceled. Several studies have been conducted on the problem of overbooking, which is used in different spheres of transport, hotel businesses, etcetera [22–25]. In [22], the study focuses on the overbooking problem in the presence of a random external environment, with a specific focus on airline overbooking.

The monograph [26] contains a collection of stochastic financial models out of which, the classical *Markowitz Problem* is of particular interest. There are n assets with random profitability's $R_1, R_2, \ldots, R_n$, and corresponding averages $r_1, r_2, \ldots, r_n$, variances $\sigma_1^2, \sigma_2^2, \ldots, \sigma_n^2$, and covariances $\sigma_{\mu,\nu}, \mu, \nu = 1, \ldots, n$. The portfolio is built out of these assets by using weighting coefficients $\omega_1, \omega_2, \ldots, \omega_n$, where ω_μ is the share of asset μ in the cost of the whole portfolio. The profitability of such portfolios is random and represented as follows:

$$F(\omega) = \omega_1 R_1 + \omega_2 R_2 + \cdots + \omega_n R_n.$$

The cumulative risk of the portfolio with the pre-assigned value of average profitability r^* can be measured using variance $DF(\omega)$. Markowitz problem is formulated in the following way: to minimize dispersion $DF(\omega)$ under constraint $EF(\omega) = r*$.

According to [27], this problem arises when a random environment exists and a reward's increment is constant for each state of the random environment, but different for various states.

Other applications of Markov modulated processes can be found in [28].

Our survey concludes by referencing the generalizations of Markov-modulated processes. Generalized counting processes in a stochastic environment are described in the study [29]. The semi-Markov process as a random environment is considered in the paper [30]. Decomposable semiregenerative processes and their applications are presented in the research [31].

References

[1] Neuts, M.F. Probability distribution of phase type. In: *Liber Amicorum* Professor Emeritus H. Florin. Department of Mathematics, University of Louvain, 1975, pp. 173–206.

[2] Kim, C.S., Dudin, A., Klimenok, V., Khramova, V., Erlang loss queueing system with butch arrivals operating in a random environment. *Computers and Operations Research*, 2009;36(3):674–697.

[3] Kim, C.S., Klimenok, V., Mushko, V., Dudin, A., The BMAP/PH/N retrial queueing system operating in Markovian random environment. *Computers and Operations Research*, 2010;37(7):1228–1237.

[4] Krishnamoorthy, A., Joshua, A.N., Vishnevsky, V., Analysis of a k–Stage Service queueing system with accessible batches of service. *Mathematics*, 2021;9,55:1–16.

[5] Asmussen, S., Ladder heights and Markov-modulated *M/G/1* queue. *Stochastic Processes Applications*, 1991;37(2):313–326.

[6] Prabhu, N.U., Zhu, Y., Markov-modulated queuing systems, *Queueing Systems Theory Applications*, 1989;5(1–3):215–245.

[7] Pacheco, A., Tang, L.C., Prabhu, N.U., *Markov-Modulated Processes & Semiregenerative Phenomena*, World Scientific, New Jersey, London, 2009.

[8] Andronov, A.M., Markov-modulated birth-death processes. *Automatic Control and Computer Sciences*, 2011;45(3):123–132.

[9] Andronov, A.M., Vishnevsky, V.M., Markov-modulated continuous time finite Markov chain as the model of hybrid wireless communication channels operation. *Automatic Control and Computer Sciences*, 2016;50(3):125–132.

[10] Dudin, A.N., Klimenok, V.I., Vishnevsky, V.M., *The Theory of Queuing Systems with Correlated Flows*, Springer, Cham, 2020.

[11] Kopocińska, I., The reliability of an element with alternating failure rate. *Zastosowania Matematki* (*Aplicationes Mathematicae*), 1984;XVIII(2):187–194.

[12] Özekici, S, Soyer, R., Reliability modeling and analysis under random environments. In: Soyer, R., Mazzuchi, T.A., Singpurwalla, N.D. (eds.), *Mathematical Reliability: An Expository Perspective.* Kluwer, Boston, MA, 2004, pp. 249–273.

[13] Singpurwalla, N.D., Survival in dynamic environments. *Statistical Science*, 1995;10(1):86–103.

[14] Andronov, A.M., Rykov, V.V., Vishnevsky, V.M., On reliability function of a parallel system with three renewable components. In: Rykov, V.V., Singpurwalla, N.D., Zubkov, A.M. (eds.), *Analytical and Computational Methods in Probability Theory*. ACMPT. Lecture Notes in Computer Science, Vol. 10684. 2017, pp. 199–209.

[15] Andronov, A.M., Vishnevsky, V.M., Reliability of two communication channels in a random environment. In: Vishnevskiy, V.M., Kozyrev, D.V. (eds.), *Distributed Computer and Communication Networks*, Vol. 919. DCCN 2018. Communication in Computer and Information Science, 2018, pp. 167–176.

[16] Chakravarthy, S.R., Krishnamoorthy, A., Ushakumari, P.V., A k-out-of-n reliability system with an unreliable server and Phase type repairs and services: The (N, T) policy. *Journal of Applied Mathematics and Stochastic Analysis*, 2001;14(4):361–380.

[17] Dimitrov, B., Rykov, V., On k-out-of-n system under full repair and arbitrary distributed repair time. In: Vishnevskiy, V.M., Samouylov, K.E., Kozyrev, D.V. (eds.), *Distributed Computer and Communication Networks*, Vol. 13144. DCCN 2021. Lecture Notes in Computer Science, 2021, pp. 323–335.

[18] Houankpo, H.G.K., Kozyrev, D., Reliability model of a homogeneous hot-standby k-out-of-n: G system. In: Vishnevskiy, V.M., Samouylov, K.E., Kozyrev, D.V. (eds.), *Distributed Computer and Communication Networks*, Vol. 13144. DCCN 2021. Lecture Notes in Computer Science, 2021, pp. 358–368.

[19] Andronov, A.M., Yurkina, T., Reliability of supply chain in a random environment. *Automatic Control and Computer Sciences*, 2015;49(6):340–346.

[20] Christopher, M., *Logistics and Supply Chain Management*, Harlow, UK, FT Prentice Hall, 2004.

[21] Wolfgang, K., Thorsten, B., *Managing Risks in Supply Chains. How to Build Reliable Collaboration in Logistics.* Berlin, Erich Schmidt Verlag, 2006.

[22] Andronov, A., Dalinger, I., Santalova, D., Problem of overbooking for a case of a random environment existence. In: Vishnevskiy, V.M., Samouylov, K.E., Kozyrev, D.V. (eds.), *Distributed Computer and Communication Networks*, Vol. 12563. DCCN 2020. Lecture Notes in Computer Science, 2020, pp. 1–13.

[23] Chatwin, R.E., Optimal control of continuous-time terminal-value birth-and-death processes and airline overbooking. *Naval Research Logistics*, 1998;43:159–168.

[24] Chatwin, R.E., Multi-period airline overbooking with a single fare class. *Operational Research*, 1998;46:805–819.

[25] Sulima, N. Probabilistic model of overbooking for an airline. *Automatic Control and Computer Sciences*, 2012;46(1):68–78.

[26] Mamon, R.S., Elliot, R.J. (eds.). *Hidden Markov models in Finance*, International Series in Operations Research & Management Science, Vol. 104. Springer, New York, 2007.

[27] Andronov, A., Jurkina, T., Markowitz. Problem for a case of random environment existence. In: Pilz, J., Rasch, D., Melas, V., Moder, K. (eds.), *Statistics and Simulation*, Vol. 231. IWS 2015. Springer Proceedings in Mathematics & Statistics, 2018, pp. 207–216.

[28] Spiridovska, N., Markov-modulated processes, their applications and big data cases: State of the art. In: Kabashkin, I. Yatskiv, I., Prentkovskis, O. (eds.), *Reliability and Statistics in Transportation and Communication.* RelStat 2019. Lecture Notes in Networks and Systems, Vol. 117, 2020, pp. 100–109.

[29] Cocco, D., Giona, M., Generalized counting processes in a stochastic environment. *Mathematics*, 2021;9,2573:1–19.

[30] Andronov, A.M., Vishnevsky, V.M., Algorithm of state stationary probability computing for continuous-time finite Markov chain modulated by semi-Markov process. In: Vishnevskiy, V.M., Kozyrev, D.V. (eds.), *Distributed Computer and Communication Networks.* DCCN 2016. Communication in Computer and Information Science, Vol. 601, 2016, pp. 167–176.

[31] Rykov, V.V., Decomposable semi-regenerative processes: Review of theory and applications to queueing and reliability systems. *RT&A*, 2021;16(2(62)):157–190.

Index

S

T

Printed in the United States
by Baker & Taylor Publisher Services

Printed in the United States
by Baker & Taylor Publisher Services